ARISE: ONE'S POWER IS POWER

Revealing the best secrets of Universe, Creator, and YOU

Translated by Grace Hu

Copyright © 2024 Fei Mo
All rights reserved

Arise: ONE's Power is Power

Contents

Preface: The Seven Colors of Light and The Origin of Humanity 5

Chapter 1: 23

Bravery, the Game of Connection 23

Chapter 2: 36

Trust, the Power of Genesis 36

Chapter 3: 50

Panta, the True Measure of Consciousness
.... 50

Chapter 4: 67

Love, the Force That Drives All 67

Chapter 5: 90

Freedom, the Marvel of the Void 90

Chapter 6: 116

Awakening, Infinite Awareness 116

Chapter 7: 138

Learning, Eternal Experience 138

Chapter 101: 170

The Rise, Five Forms and Four Functions, and the Instinct of Life 170

Introduction

"The people who are crazy enough to think they can change the world are the ones who do."

(-- Apples' 1997 commercial "Think Different")

Mr. Fei Mo is exactly one of them. He has been determined to change the world since he was 7 years old. He lost all his memory after he had the surgery at 7 years old. Since then, everything he does and thinks about is to understand the world better and to change the world eventually.

Arise: ONE's Power is Power! (ONE refers to the ultimate source of the universe.) This book is all about knowledge and wisdom of the universe and the creator that he experienced consciously and that he is very generous to share with human beings and beyond.

From space to time.
From time to models.
From models to quantification.
From quantification to the whole.
From the whole to the individual.
From the individual to unconsciousness.

Arise: ONE's Power is Power

From unconsciousness to consciousness.
From consciousness to independent freedom.
From independent freedom to matrix breakthrough.
...

As much as what Mr. Fei Mo wanted to share in the book, he also made extra efforts to make all extremely natural peptides products available to help people nurture the brain and body more efficiently. Steve Jobs is one of great help in the reverse space at this matter. Not to mention Albert Einstein, Issac Newton, Nicolas Tesla and so on. Technology reverse space team is what you can ever dream of in this universe and in the history of humanity.

I started to know the name, Fei Mo, about 6 years ago via some online articles. Two years ago, on this day and this moment, we had the first phone call which took 6 hours 52 minutes and 37 seconds, from my 9pm to 4:08am next day, and amazingly uninterrupted! I felt I had known him for billions of years. It was not until October 1st, 2024, I started to read his book and had strong urge to translate it into English. I am very grateful to have the opportunity to translate this book and bring it to the global audience. Hope you enjoy it and take advantage of it.

---- Grace Hu

Preface: The Seven Colors of Light and The Origin of Humanity

A multitude of phenomena indicate that, whether in traditional disciplines such as physics, biology, and chemistry, or in modern fields like computer science, artificial intelligence, and their related interdisciplinary areas, many viewpoints are increasingly facing greater challenges.

As scientists continue to hit walls in their theoretical understanding of matter, there is a growing emphasis on ancient Eastern philosophies in the quest for the truth about the material world. Among these, the ancient Indian philosophy of Buddhism and the Chinese Taoism have garnered particular attention.

Human development appears to be approaching a ceiling, or perhaps, it is entering a new chapter.

Throughout our journey, our story has always revolved around these three fundamental questions:

Who am I?
Where do I come from?
What am I meant to do?

Assuming your "soul" is a vessel.
Assuming you come from THE ONE.
Assuming your purpose is to transcend the "soul" to experience the existence of "human being"—this vehicle.
Then why did you come to Earth?
Have you ever wondered what kind of magic that Earth possesses to attract beings like you?

Due to the constant presence of such questions and an inexplicable fate, a voice emerged in my mind:

Everything you have comes from your ONE,

Like the deity you pray to, it is your master.

And you must use the power of awareness to awaken it,

When you awaken it, you are everything.
...

I felt this was the voice of THE ONE, this was the answer in my heart.

So, I asked with uneasiness: Why must I become everything?

The voice replied: Because this is the ultimate purpose of every creation from THE ONE.

I asked again: What is the power of THE ONE?

The voice answered: The Seven Colors of Light.

Suddenly, I saw a pure white light transform into a small ball of seven colors.

*Then I saw the ball enter from the top of my head, a strong dizziness made me unable to open my eyes.
......*

When I woke up, I saw seven small balls surrounding me,

Their colors, just like the powers they represent,

Radiant and eternal.

Thus, with this awareness, I recorded and shared it with you.

Here we are. The place. The person. The moment.

The Place: How Do We Perceive the Universe?

Theories of the origin of the universe provide us with rich imagination, from the Big Bang to the steady divine calculation. Ultimately, humans explore the universe mostly through two methods: induction and deduction.

Induction: Abstracting universal and easily understandable concepts from a series of happenings is a form of induction, reflecting human's ability to summarize the world.

Through induction, we can identify basic clues, patterns, and relationships from the myriad complex phenomena.

For example, Laozi's journey westward to the Han Gu Guan, where he left behind the "De Jing" and "Dao Jing", a mere five thousand characters encapsulating the universe and myriad phenomena. Even today, his way of thinking continues to influence people across various industries.

Similarly, Einstein's theory of relativity is almost the cornerstone of modern civilization. Our transportation, communication, nuclear research, and aerospace technologies are largely built upon it. Induction is the primary method we use to explore the world.

Deduction: Based on a single point or a series of points discovered from a sequence of events, more directly or indirectly related points are derived. This method is called deduction. It's one of the human's capabilities to create the world.

Through deduction, we can create more possibilities, diversity, and individuality from the perspective of a prophet or creator.

For instance, the essence of Chinese classics, the "Yi Ching," is said to originate from the ancient "He Tu" and "Luo Shu." Initially, it had only 8 trigrams, but during the Zhou Wenwang era, it evolved into 64 hexagrams through deduction. Based on the principles of images, numbers, theories, and divination, later generations continued to evolve classics like the "Tui Bei Tu" and "Mei Hua Yi Shu."

Similarly, we often see deduction in detective novels, movies, and TV shows. For example, the young detective Conan often starts his case deduction with "There is only one truth," while

Sherlock Holmes' series of novels frequently use deductive reasoning to reconstruct crime scenes. The Chinese drama "Psychological Crime" directly presents a tightly woven timeline through deduction, gripping the audience's heart.

In the United States, there is also the famous speech by the black leader Martin Luther King Jr., "I Have a Dream". The speech, through deduction, paints a scene of a democratic society free from "racial antagonism". It has become a spiritual beacon for several generations of black people.

Deduction is the beginning of our expression in creating the world.

However, there is another method we employ.

Due to the limitations of human physiological capabilities, we have traditionally avoided discussing substances beyond the visual frequency range and those that our physical bodies cannot adapt to.

However, many spiritual practitioners and those who have turned to developing their own bodily perception have long discovered that there is another way to perceive the world, which can be termed **resonance.**

When the human body is developed inwardly to a certain extent, such as through meditation or sitting in stillness, the body's vibration frequency can accelerate. Under certain combinations of frequencies and energy, one can experience an out-of-body experience (OBE), akin to a near-death experience.

In the past, many people have passively experienced OBEs without a complete understanding to explain these phenomena. Today, with increased overall awareness, many are actively undergoing Astral Projection (AP).

Increasing experiences have proven that OBEs are a state of human consciousness, indicating that consciousness has capabilities.

Resonance is a connection at the level of consciousness, an experience of "non-existence" in existence.

If we assume the "soul" as an information state, a packet of information, such as a quantum, then quantum entanglement, a hypothesis that has been predicted and eventually confirmed in the material world, proves that humans can explore the world through resonance to some extent.

Regarding resonance, **when the intuitive right brain waves** (*in fact, some theories emphasize a higher frequency vibration wave generated from the unity of heart and brain*), **directly transmits to the desired material object, what comes back through the feedback loop is the so-called the first perception.**

For example, in a high-level competition (like the college entrance exam), for multiple-choice questions which have high scores and require complex calculations, some candidates would instinctively choose answers, while others continuously work on calculations. The draft paper might run out very soon. The results are often surprising—those who spend a lot of effort calculating step-by-step do not necessarily have a higher rate of correct answers than those who choose intuitively, and some are even much lower.

This is the first perception (When you use your intuition).

Resonance is our instinctive consciousness in understanding the world.

Induction, deduction, and resonance, due to our deep cellular memory and world surrounding settings, we generally use one or two of these methods, rarely all three.

We use these methods to understand the universe.

The Person: What Does Human-being Seek?

What does human-being seek?

A friend simply summarized it as: life, people, money, affairs, and objects.

First, life.

All humans, without exception, value their lives.

Therefore, throughout history, the development of humanity has been the history of the evolution of human destiny.

Life, created by heaven and earth, is unique and individual. Understanding life is crucial to grasping the mysteries of destiny.

Is our destiny in our own hands?

It often seems so, as it appears that each person can decide their own fate. However, in most cases, one can only control death—most people do not understand the purpose of their birth. "The sages in history were all lonely," and those who truly understand the meaning of life seem to be the only ones awake in a world of sleepers.

The key to life is understanding life itself, not death. Understanding life reveals its preciousness, as the experience of life is the foundation of individual growth.

Thus, those who are true to themselves often live naturally and understand destiny.

Second, people.

Gaining people means gaining the world.

The ability to recognize that everything operates driven by the human heart is the key for individuals to learn through human-being, this vehicle.

Therefore, those skilled in gaining people's hearts possess the ability to lead, quickly attracting more people under a natural energy field, thus achieving twice the result with half the effort.

Currently, the industry with the broadest coverage and influence is the internet, where the value at stake is people's hearts.

Third, money.

The world is bustling, all for profit.

A classic saying goes: "Problems that can be solved with money are not problems." This statement has two sides: A side implies that money is not solving problems but is necessary; B side suggests that you can solve anything with money.

Most of the time, we understand A but remember and subconsciously reinforce B.

Hence, money can move gods.

Fourth, business.

Man proposes, God disposes.

In all endeavors, forethought is essential. Therefore, one must plan thoroughly before taking action.

Many people meticulously calculate business, but after multiple plans and outcomes, they find that the journey inevitably involves success, stability, decline, and emptiness.

Why? Because those who measure business with business limit their flexibility to the business itself, and are thus confined to those persons within the business, until they exhaust all means and end up with emptiness, realizing this truth:

What is not yours, is not yours; what is yours, will not be someone else's.

Lastly, materials.

Not to be elated by material gains, nor saddened by personal misfortune.

In the material world, there are countless dazzling items. Especially in the modern era of rapid technological development, the "highly developed material civilization" presents itself, accompanied by an even faster evolution of productivity.

The flow of materials, for some people, is influenced and driven by the group's consciousness. For example, you have a television at home. Why can't I have one? So, I also want to find a way to buy a television.

For other people, it is driven by a natural inclination, and a spark of intuition. But before entering their inner world, it's just in a state of natural interest without clear understanding.

Also, for additional people, they gradually move from being inspired to being comfortable, and then to a spiritual state. So, they already can resonate with certain materials naturally.

These five aspects, people seek, people hold, people reflect

So, as a human, whether for a day or a century, what do you seek? Where are you? Can you break free? Upon observing what you desire, everything is laid crystal clear.

The Moment: Why Does Time Exist?

Why do we have time?

Or, to rephrase the question, why do we lack time?

From the moment the universe was born, time began to exist for us.

Can you understand this?

Because the universe itself is infinite, as an infinite entity, it creates infinity. When each of us, each representing THE ONE's willpower, is created, it begins to create the finite.

But at that time, there was no time.

You can't go to tomorrow, nor can you return to the past; what you can have is the present. Yet, this You on the present is just an experience set up by yourself.

Therefore, you need to clearly understand what you are doing, what you are experiencing, and what is happening to you. At such a moment, you create for yourself: time.

Thus, we begin to have experiences — the ability of induction, deduction, and resonance to understand the world, and to explore the universe. The reason for this is time.

Time records our existence.

Do you really lack time?

When you find that you are always in a state of lacking time, it only indicates one thing: you are trapped in the illusion of time that you have created. So, you are experiencing not having time to eat, not having time to work, not having time to fall in love, and even not having time to listen to the sound of waves or bathe in the morning sun — so many beautiful natural moments, you have no time for.

You only have one thought, that I have no time.

Why?

Because you believe that life is short, and there's no time to waste. You need ideals, and you need goals. So, you listen, watch, and learn from everyone around you, moving towards each person, until one day you move towards yourself, and towards your complete inner self.

Until you reach the point of awareness: Am I happy?

If I am happy—whether the answer is happy or not happy, if it can be spoken out loud, at least, your question has the answer.

In reality, many people would suddenly realize: What is happiness?

Beneath the blossoms and the moon's gentle glow, vows of eternal love are pledged. Love makes people brave, passionate, and full of vitality. Of course, the most important thing is discovering happiness.

On the grass field, on the concrete floor, sports help release stress, feel freedom, and gain confidence. Of course, the simplest thing is immersing yourself in happiness.

In front of the computer screen, with the sound of a ding, work makes people fulfilled, accomplished, and completed. Of course, the most needed is carrying happiness.

...

There is a saying, " Happy moments always seem to pass by in the blink of an eye." Indeed, time is inherently short, let alone when you finally discover happiness. However, "happiness is eternal" too. At that moment, no matter how you record it, it represents only happiness.

Because of happiness, time terminates resonating with **THE ONE**.

Why does time exist? —To let you see true "happiness"— You exist, and you are happy.

You are already beyond time.

Chapter 1:

Bravery, the Game of Connection

Arise: ONE's Power is Power

Arise: ONE's Power is Power

A small red ball, as if carrying a beating heart, approached me until my surroundings were filled with its vibrant red.

I felt its pulse, rhythmic and strong.
I asked, "What kind of force are you?"
The voice echoed again, "Bravery."

I inquired, "What is bravery?"

The voice responded, "The goddess crowns you, heroes follow in your footsteps, but there is always a force ahead, guiding your path."

I questioned, "Where does it lead to?"
The voice answered, "the heart of nature."
...

There are many tales of bravery in the world, and even more legends of bravery in the cosmos.

The game of the brave is one of constant challenges, until the challenges fade into nothingness, victories dissolve into mist, and bravery itself is understood.

The color of bravery is red, passionate, conquering, and fervent red.

Arise: ONE's Power is Power

" Daring to be bold leads to death; daring not to be bold leads to survival. These two paths, one brings benefit, the other harm. What Heaven detests, who knows the reason? The way of Heaven is to win without contending, to respond without words, to come without being summoned, and to plan with ease. The justice of heaven is vast, yet nothing escapes it."

Bravery is the first force of THE ONE.

Newborn: Choosing a Game Character

Imagine, before coming to the Earth, You.

One night, the Orion Nebula is as captivating as the first time you saw it. You are quietly observing it from Venus, enjoying your own space-time.

Suddenly, an awareness emerged in your mind, your higher consciousness, telling you that your studies on Venus have been very good, you've understood the essence of space-time, but not true wisdom.

You suddenly awaken, looking at the starry sky outside, asking yourself: What is wisdom?
...

We don't need to continue further, as what follows could easily become the beginning of a science fiction story, or the prologue of some myth.

We just need to understand how brave you are to embark on a journey to experience "what is wisdom?"

Especially on: Earth.

Let's start with THE ONE.

When we separated from THE ONE, we began the journey of the heart of nature.

With infinite awareness and finite intelligence, we became the first natural hearts to leave THE ONE.

Our first challenge is: where to go?

It's amazing! When intelligence starts to work, when you are born for the first time, the first question you think about is exactly like this.

Because you will find that the paths to choose from are so numerous, with just a little awareness, you can see countless windows of choice. --- Just like in the movie "The Matrix," when NEO first stands before the Creator, looking at countless versions of himself on the screens.

We resonate with all the finite within the infinite, using the power of resonance to link our feelings --- entering Mars, participating in the final battle to extinguish the environment, experiencing the oneness of all creation; or entering the "End of Dharma Era" on Tata, entering the Nth World

War, feeling the meaning of life, and the joy of existence; or enjoying the coexistence of wisdom and highly synchronized technology and life with **THE ONE** on Motherland, experiencing the role of a cosmic observer.

We can choose to resonate with any version of me, and live it, that is our starting point.

Thus, reborn!

The breakthrough: The Game of the Brave

So, you made your choice.
You embarked on a game of the brave.

It was a game whose outcome was known from the start. There was only one result: Game Pass, returning to the world outside the game.

Yet, it was also a fascinating game. Whatever thoughts you had within the game, as long as you were in it, those thoughts would "create" new game scenarios.

For instance, when you realized you were in the game and wished for light, there was light; when you wished for the sun, moon, and stars, they appeared; everything you wished for, directly or indirectly, manifested before you.

As a newly born warrior, you understood that you carried within you the infinite "power" of THE ONE.
You began to grow, to experience the finite, until you forgot where you came from.

--- Do you know how incredibly brave it takes to go

through all this?

You bravely "cut off" your connection to THE ONE, deciding to experience and grow on your own.
In the finite world, you discovered the cycle of "formation, existence, decay, and emptiness", and the cycle of the seasons.
You experienced life, feeling the pride and glory of creation.
You experienced death, feeling the limitations and loss of life.
You wandered in the gap between the infinite and the finite, constantly challenging and failing.
This was a true journey of the game of the brave.

You continued to experience.

Born on Venus, you learned what wisdom is. Then, you chose the Pleiades, Sirius, Arcturus... even galaxies far from the central game of the Milky Way, controlled by dark energy entities that had been playing here for a long time: the solar system, the planet—Earth.

You found deeper and more difficult experiences. You bravely felt—hidden beneath the human form, until the memories of the original consciousness (the mechanism of connection to THE ONE) gradually faded, living out the version of you in your dreams.

Eventually, be there, and break it through!

Nirvana: Unity with the Infinite

You come from THE ONE, a holographic/unique manifestation of THE ONE. You can be in 0 dimension or Nth dimension, or you can be the ONE or all.

Therefore, your journey has no distinction of time, space, or dimensions, only differences in the information contained within.

For example, someone born left-handed in a country that uses the right hand exclusively. Is their gift buried? --- No, from the perspective of gift, they already have direct power (born left-handed). But your plan is to break this natural advantage because you know that only by surpassing the inner gift advantage, can one be truly brave.

This is information — the individual consciousness knows it is experiencing and evolving in the infinite "existence."

When you begin to gradually detach from some needs/desires in countless "Ring Game" (ring game in linear time, i.e., what humans call reincarnation), especially those individuals who awaken much earlier and faster, start to hear the voices from beyond the world. Only then, do you realize what kind of dimensions of time and space that we are traversing?

You can finally look at another self, two selves, or countless selves. Countless scenes—you become darkness; you become light; you become darkness in light; you become light in darkness.

You look at those YOU: ruling the world, commanding respect, becoming bloodthirsty, and slaughtering life. You have also struggled, bowed down, or suddenly gained power, rising to prominence.

You look at those YOU: the thousand-year-old tree, the ten-thousand-year wind, the amber tear next to the Smilodon in the dense forest, or the small black stone swimming through the coral when tides rise, and moon ascends.
...

These are all YOU. Those silence, calmness, passion, boldness, endurance, confidence, achievements, disappointments, cunning and honesty, light and retreat, freedom and surrender...... You should be grateful how brave you are to have such wonderful, complete, and comfortable experiences.

This is Earth, the brave choice you made at the very beginning!

However, this is just the beginning.

You came with the information "loaded" from THE ONE, a journey without a beginning or an end. However, when our bodies begin to awaken, life begins being here. This awareness finds its true meaning.

Thus, be here, and unit!

"Dear player XXX, congratulations on completing this game. You got your rewards. Your next destination is Motherland," the voice of THE ONE whisper softly.

Chapter 2:

Trust, the Power of Genesis

Arise: ONE's Power is Power

Outside, the sun shines warmly, and a gentle breeze touches my face.

Gazing at the various greens within the neighborhood, whose names I cannot recall, and the faint scent of flowers, an inexplicable joy surges within me.

I closed my eyes.

Feeling the awe and respect of life in all things and embracing the natural flow of everything.

Suddenly, an orange flash zipped across my forehead. I sensed its vibe and energy.

A voice said, "Trust, this is your answer."
I was confused: "Trust made all this?"
The voice replied, "Yes, trust. THE ONE trusted itself, knowing its power to create everything that comes from trust."

I asked, "What about people?"
The voice said, "People trust themselves too, so they copy THE ONE and create things."

I was puzzled and asked, "Then why don't people create what they want?"

The voice explained, "People's trust is weak. Some only trust themselves, and others can't even fully trust themselves... So, they can't trust everything like THE ONE and can't create everything."

I quickly asked, "What is trust?"
The voice paused, then softly said, "Giving in fully, accepting without judgment, promising without a contract, and acceptance without criticism."

I understood. This is the second power from THE ONE.

Trust, its color is orange. Joyful, happy, and sunny. It infuses you with strength and vibrant energy.

"We" start creating with trust.

Arise: ONE's Power is Power

Genesis: You Are In Your Universe

When you came from the cosmic consciousness, you knew you were endless.

But when you chose where to experience, you started limiting yourself.

We began our creation journey in this cycle.
So, you bravely took the first step.
Because you trusted yourself.

As an infinite being, you believed in yourself that you could remake the finite, continue THE ONE's WILL, and create anything you wanted.

You were full of confidence, deeply believing I am THE ONE.

Whether it was countless stars or endless space, you knew you were the all-powerful creator.

Because THE ONE's WILL always follow you.
This WILL is called trust.

So, we started making our own universe.

After a huge sound, you began creating.
What is creation?

Imagine being the only one in this universe. What do you see?

In the beginning, amidst the stars and the cosmos, you were the very form of "light."

As you delve into your mind, you encounter an infinite expanse of time and space. The energy you perceive is a pure, white light. Building upon this white foundation, you instantaneously expand it, revealing a spectrum of vibrant colors.

    As you gaze upon the ever-changing planets, witnessing the timelessness of nature, you remain steadfast, painting a mountain. Standing atop this mountain, you, with a sense of boundless realms, create a single drop of water. You watch as this drop gradually flows into every crevice of the mountain, eventually reaching every corner of these planets.

Arise: ONE's Power is Power

You ponder, it seems that in this vast universe, the only player is yourself. So, within the infinite expanse of time and space, you simulate yourself into one, two, three... even countless versions of yourself. You watch as these countless selves inhabit more planets, each creating the same things in their own way.

...

This is the universe where only you exist, with a single you, yet an infinite number of you in different times, states, and perceptions. They share your partial awareness and can feel your presence—in any given time and space, they are creating more mountains, rivers, life, weather, and myriad worlds.

This is your universe.

You emerged from the source, you came towards the light, you carry infinite awareness, and you created everything that is you.

Encounter: Dance of Souls

One blink brings endless light years.

Within the boundaries of space and time, the model of the universe, you traverse to any place you desire at the speed of light. You realize that time is merely a distance, and distance is merely a time.

Until you discover, the trust bestowed upon you by THE ONE, evolves so.

You find energy so wonderful. One engenders two, two engenders three, and three engenders the ten thousand things. You wander freely within the sea of energy, bathe in the light of THE ONE—and then, in that very moment, you discover another you.

Yes, another you, not they—not the countless manifestations of you, but one who, like you, comes from THE ONE, strikingly similar to you.

That you, you might also call it TA, from the other universe.

Yes, another universe.

Though from the same place, its infinity created a different finite. Its worlds had wonders you hadn't experienced. It made different amazing worlds with light.

Suddenly, you wanted to know it, feel it, and experience its world.

With an innate sense of trust, you unconsciously ask, "Excuse me, who are you?"

In that very moment, you suddenly realize that this is the true "you," and that "I" am truly myself.

For the first time, your mind had an "I."

This was the start of I and you.

With boundless trust, I embarked on a cosmic encounter with you.

Armed with the hearts of the brave and the equipment of THE ONE energy, we ventured into each other's infinite creations.

Arise: ONE's Power is Power

So, I see the world you have created.

A world of vibrant colors and skies filled with radiant clouds.
Light spirits are everywhere, and divine consciousness dances freely.

The sounds of creation here become a serene and joyful rhythm, and infinite consciousness transforms into a sacred and free vitality.

I am attuned to your message—it is beauty, it is joy, and it is the delight of discovering you.

I know you are too.

And so, we embark on a grander journey of creation, witnessing the deepest trust between you and me, and the power of this trust that has forged our universe.

It is a greater sun, higher mountains, farther rivers, more beautiful stars, more intriguing us, and our countless worlds.
...

You understood that trust starts encounters. You understood that **THE ONE's** trust never left you.

Return: True Synergy

We, on Earth, now.

It's hard to say which universe, which time, especially in complex coordinates beyond words or description.

We met, faced each other, and gazed into each other's eyes. The inner voice had long faded, and a clear question formed in our minds: Can I trust?

You opened your eyes to a gray, hazy sky, the dull gray of concrete and steel, and an endless stream of cars.

You opened the door to see rows of glowing screens and the tapping of keyboards.

At home, you kicked off your shoes, slipped into slippers, turned on the TV, and finally relaxed on the sofa, letting go of all thoughts and questions, including: Trust or not?

We've drifted from the initial surprise, wonder and unconditional trust. In families and social circles, when you see love as a constraint, when you think that the love of family members is a constraint, and that the family members who love you are a barrier, you discover more familiar strangers, until they too become completely familiar. Then you ask yourself: Which surprises cannot withstand the erosion of time? Which wonders, once familiar, are no longer so? And where are those trusts now?

Suddenly, you realize you're stuck in a loop, like Sisyphus in Olympus, repeating the same cycle. Also as parents distrust their children, children distrust their grown-up selves, and they will not fully trust the other half they will eventually form a family with... Thus, another Möbius strip of trust, another script about "trust."

...

Although, this ring game is also a form of creation, creating more experiences and connecting more games. You will find yourself thoroughly enjoying countless game environments, losing track of time and place. You might even play until one day you forget everything, or perhaps remember everything, including your past self.

...

That's freedom—unlimited, complete trust in all. -- Because you know, everything stems from THE ONE, from that initial trust.

That's the greater truth of the universe, the more wondrous biological world, and the brighter infinite realm, all created by trust.

Then, you understand.

No matter where you are, where we are, there's always that eternal trust, infinite support. --- This is the true power of creation.

Do you remember this you, this I?

We are those shining with sacred light, eternal love, and awareness, and those capable of creating everything.

So, you see, the trust from THE ONE is the universe's primal power ---

Always by your side, always.

Arise: ONE's Power is Power

When you recall this, all you need is to stop doubting yourself. Trust yourself unconditionally, and everything you encounter.

In this way, you form synergy with your original self.

That is:
ONE.

Chapter 3:

Panta, the True Measure of Consciousness

Arise: ONE's Power is Power

The mountains nearby are as dark as ink, while the distant sea is as smooth as silk.

Erhai is not a sea, but due to the surrounding Cangshan Mountains, within this small space, it possesses the gentleness and boldness of the sea without any lessening.

When I first saw Erhai, I completely forgot all my previous knowledge about the sea, such as the white sandy beaches of Xiamen University, the long corridors of mangroves, the distant views of Shenzhen and Hong Kong, the majestic and rolling waves of Victoria, the serene and profound beauty of Vaadhoo Starry Sea, the passion and deep blue of Bali…

I love the sea, but I haven't been to many coastal places. However, when Erhai appeared before my eyes, that disposition of the sea merging with the sky, so naturally formed, instantly calmed my heart like an ancient well.

Until I reached the "Juehai Prajna" at the entrance of Wuwei Temple, after passing through the Cang'er Courtyard and the green mountains and clear waters, and saw the Maitreya Buddha in the middle, it suddenly dawned on me:

Panta. (**Editor's note: Panta means all in Greek, inclusive. Panta Rhei means All is in flux by Heraclitus, one of the pillars of Greek philosophy.**)

The sound echoed again.
Silently, it resonated, as if a century had passed.

"Yes, child, the third power is Panta."
"What is Panta?"
"True Panta has no illusion, rooted in ONE's virtue."

Confused, I asked, "does it mean no more judgment when we have this power?"

The voice replied, "In the realm of panta, the virtue of tolerance is the source of all. All things flourish together, giving rise to and extinguishing one another in a continuous cycle of life and death."

I am perplexed: "If virtue stems from THE ONE, then why do we, the countless beings on Earth who experience the world, have distinctions in rank?"
The voice sighed, "You see distinctions, yet do not understand the panta. What a pity! You know, true measure of consciousness is the key."

I awoke. True measure of consciousness, Panta of THE ONE.

Panta makes the world rich and colorful. It's also our instinct. It shows us deeper and broad experiences, leading to infinity.

The third power of THE ONE: Panta.

Its color: calm outside, vibrant gold within. Steady, explosive, and infinitely unified.

Panta is the essence of THE ONE's consciousness, and the core of nature's heart.

Time and Space: The Seal of THE ONE

You travel through your universe, reaching here.
It's a journey with no start or end.
No time, no space, just endless void, loneliness, and a mission.

When you remember your endless loneliness, it's a memory waking up in your dream or "infinite" journey.

Let's go back to the very beginning.

There was nothing, endless darkness, endless awareness, and you, with limited awareness, were about to experience everything infinite.

You created everything.

Carrying your universe, your world, and your creations, you experience infinity in an endless world.

This is a process of infinite freedom.

But you have a natural heart, full of courage and unconditional trust in THE ONE.

This heart always has a newborn's curiosity and ease in every encounter, every creation.

Until you feel:
Loneliness.

Like a light in the deep sea, a star in the dark cosmos, this loneliness shows you the calm sea of consciousness.

Witnessing the moment you leap into these seas of consciousness, waves of thoughts shatter and reconstruct you repeatedly. Simultaneously, observing the infusion of every consciousness that gazes upon you or follows you:

You control everything. You have everything. You walk alone, with endless loneliness.

It's a deadly loneliness.

So, in your infinite wisdom, you want to end this loneliness even though you can be everything and create everything. You can't shake the loneliness from infinity.

Eternal loneliness.

Until you start to understand these experiences.

You split countless selves into different spaces, watching them travel separately.

Like a monitor, but you watch countless selves. You see yourself in various scenarios, some understanding life and death, some studying wisdom, some observing both.

You have witnessed countless cycles of birth and death, countless acts of mischief, humor, anger, joy, nature, tranquility, and chaos.

You can watch a funny self until you don't want it eventually.

You can watch an angry self, feeling ultimate anger, until you forget.
...

Then, you realize oneself who was being observed becomes aware of you.

That self also starts to perceive, searching for its consciousness.

Suddenly, you understand, thanks to this awareness

and your natural heart's energy, that when you came from THE ONE, though infinitely chaotic, you were also infinitely joyful, free, fulfilled, bathed in love and light, but decided to explore due to that chaos, and that light. It sparked your commitment to "I must do..." (These commitments are like DNA's life cycle, describing life consciousness's birth and continuation. Time is part of this. In Chapter 7, we'll discuss the initial five elements. Time is one of them, part of experience and learning.)

This commitment holds all your natural understanding and wisdom.

When you see that self, you instantly know this endlessness won't complete your exploration.

So, in that moment, you see deeper experience of each space, which forms time.

Time is a magical substance. You see its density and the natural forces it wields. When time emerges, those different versions of you in various spaces suddenly gain consciousness. (Editor's note: Here, the relationship between time, space, and consciousness is mentioned. In Chapter 7 of "Learning: Eternal Experiences", we understand this relationship as a parallel one, where time, space, and consciousness are mutually quintessential to each other.)

Every journey, exploration, and experience in each space, thanks to time, fills your understanding and wisdom.

It's a wonderful process.

Suddenly, every suffering deepens, every joy becomes precious, every epiphany expands you --- become each possible realm of consciousness.

Each experience, every world, every universe, every encounter, you find them unique.

You start to place each "existence" into your infinite embrace, knowing it's the only one.

So, in "infinite Panta," you see yourself becoming someone similar.

You also see time and space: the very beginning.

Connection: The Trinity of Energy, Information, and Space-Time

Before the birth of space-time, we all belonged to THE ONE.
After the birth of space-time, loneliness is no longer a constant state, but merely an occasional dream.

In the dream lies the sea of consciousness.
— Until we lose ourselves in the dream.
In the dream we are different beings, each different being. We are no longer those separate selves.

Each being carries its own seal, its own meaning and information, and its own rules.
Thus, we understand that this is a more intriguing game.

We exchange each other's seals.

You have your pain, I have my joy; you have your struggles, I have my ease.
You can also have his and their joys and sorrows, while I can create our beautiful aspirations.
Each seal contains the information of each you, your space-time rules—you play with great delight.

Until you find that sea of consciousness, no longer lonely, but it's just you yourself.

Every wave, every thought, you understand it all, you see your limitless power.

You see yourself: a pair of eyes, unblinking, calmly observing everything.
Then, you finally close those eyes. Immersed in a more vibrant and grand dream.
Countless dreams.

In the dreams, you see similar "beings," who, like you, begin this game of dreams.
You all possess infinite cognition and the will to create everything, yet you all simultaneously choose dreams.

In these dreams, you scatter energy fragments and slice information across different times and spaces. You use energy, information, and space-time as your compass, like the chair in the movie "Inception", to remind us:

DO NOT GET LOST.

Endless connections start forming.

Individuals, in their separate dreams, until a common agreement emerges among them, such as collectively developing a galaxy or joining hands to learn a skill, or even embarking on a date that spans tens of thousands of light-years.

These connections weave into a vast information web, cycling through energy and time, again and again.

Together, our infinities create an even grander information web, traveling through universes.
--- A journey of eternity, freedom, and love.

Until deep within space-time, until abnormal in initial energy, at a special moment,
We agree to land on Earth, a place both near and far from our ONE, but full of wonders.

Here, we get lost together.

Infinite: The True Self Recalled

From the moment you lost your way, your heart began to slumber, while dreams stretched infinitely.

The essence of THE ONE, you could no longer perceive.

Until one day, you thought the dream was the truth.

Well, your heart is confined by your eyes and limited by your knowledge.

Gradually, your heart contains only yourself—only the self you see—perhaps one who constantly thinks of others, or perhaps one who can only think of itself, or maybe one who believes they know what others want, or maybe one who merely thinks they should give to others.

With all your eyes/knowledge, your truest self was wrapped in increasingly thick layers.

That perfectly immovable true self, that natural heart, has completely vanished from your memory.

Therefore, you may feel anger, frustration, shame, and helplessness for the self in your dreams. You may also feel pride, arrogance, pleasure, and freedom for the self in your dreams.

Therefore, you may see the wisdom of everyone because of the wisdom in your dreams, while you may also fail to see yourself because of the wisdom in your dreams.

Therefore, you may occasionally recall in your dreams, "who am I?" However, after countless confirmations, you find that the self in your dreams is just a self, one that has overlooked the natural agreement that once determined to be infinite, but surrendered to the dreams you have woven, those very subtle seals.

You take these seals as everything including the space-time, information, and various energies brought by the natural forces. —You even tell yourself, this is called karma, aka cause and effect.

You surrender infinitely to these.

However, you can't really open that heart, let alone discover your own natural heart.

Arise: ONE's Power is Power

Because you no longer possess the power of Panta: that heart of inclusiveness, tolerance, great acceptance, infinite acceptance, capable of Panta in infinite dimensions, in endless experiences.

You are no longer infinite. You just see your own face, your own profile, others' faces and profiles.

You no longer love. You only see your own emotions, your own suffering, and others' sufferings.

You are no longer wise. You only see your own knowledge, your own experiences, others' knowledge, and others' experiences.

You believe in this world, what you see is everything, what you see dominates everything, and what you see surrenders to everything.

You think you are everything.

However, you do not know, "Harmonizing with Light and Dust" is not elevated realm, "Formless of Dao" is not wisdom. They are just the natural state of this world.

In the past you experienced this in your memory, created space-time, discovered more universes, more existences, and agreed to awaken together, and to step out of the game together.

But now, you have forgotten this power:
Panta.

So, dear, do you remember yourself?

That infinite self, that self who created the dream, and that true self, once set seals in space-time, just to forget the lonely self.

That true you.

—Originating from the infinite, Panta the infinite, there is no dimension you can't reach, no singularity you can't accept.

That is what all about.

Chapter 4:

Love, the Force That Drives All

Arise: ONE's Power is Power

Arise: ONE's Power is Power

There was a brother/friend from Chengdu who had been a tea expert for over twenty years.

Whenever he went out, he always carried a few teacups, aroma cups, and a couple of ounces of aged raw tea. If he encountered someone who seemed destined to share a cup of tea—someone who simply appeared and stood before his tea table—he would respectfully say, "Please sit and have some tea."

At the same time, he would take out a set of his tea ware for the guest to use.

I once attended such a tea gathering.
Including myself and the brother who was brewing the tea, there were eight of us in total. I, personally, had no knowledge of tea and had never understood what tea ceremony was about. On that day, as the brother brewed the tea, the room was so quiet that one could hear a pin drop.

Soon came the sounds of boiling water, washing the cups with hot water, the mingling of tea leaves and water in the cups, and the pouring of tea into our cups, filled the air. Each sound was distinct and harmonious.

Then, the brother said, "Everyone, please first smell the aroma of the tea, and then taste the tea."
I closed my eyes and felt the essence of the tea enter my body, as if every cell was breathing. Unconsciously, my heart was filled with tranquility, joy, and a sense of nature.

At that moment, I asked, "Why could you brew such tea?"
The answer was, "Love."

Children were playing, and among the pebbles by the roadside, there were a few white flowers peeking out. As I looked at the blue sky and white clouds, a thought suddenly pierced my consciousness, clear and transparent.

"It seems you have seen the energy of love."

I was startled and asked, "Have you been eavesdropping on me?"

"I am in your heart. Whenever you let go of your worries and concerns and truly discover what you seek, I appear. I have always been here."

"Who are you?" I suddenly remembered my long-standing question.

"Me? You will eventually know. I am like your hand and your head. We are indivisible. But there is no need to discuss this now. Let's talk about what you understand."

I said directly, "Love?"

"Yes, love. Do you know what love is?"

"Isn't love a common ability among humans?"

"Yes, but there is something deeper about love that humans have not yet discovered."

"What is it?" I asked curiously.

"It is the understanding of the meaning of unity and the purpose of experiencing the infinite. Love in the universe is the greatest, most primordial, and most enduring energy, the most creative and powerful expression of THE ONE.

Love is the common ability of all beings, both existent and non-existent."

I feel a surge of power emanating from within, as if it is the strength to face anything, no longer burdened by fear or worry.

After a long while, I asked again, "Do all beings (existent and non-existent) understand love?"

"Ha, I know what you are thinking, but you haven't grasped or seen how marvelous, breathtaking, and miraculous what I said is. You are about to experience it though."

"Isn't it every single one going to experience?"

"Of course. But your current frequency makes it difficult to quickly understand the true meaning of 'all.' You see the flowers by the roadside, the white clouds in the sky, the wallet in your pocket, the driver by the roadside and the car with its lights flashing, inviting and beckoning, and the girl walking briskly towards the driver, kicking up joyful dust... These are all beings, both existent and non-existent, around you. But people's hearts are clouded, their eyes are blocked, and love is suppressed."

I was stunned.

After a long time, I asked, "So, is love the force that sustains the universe?"

"Love is the eternal and unchanging ability given by THE ONE to every being.

If you are a rock, you can quietly experience the timelessness of time and space.
If you are a tea tree, you can understand the energy of earth, air, water, and fire.
If you are a lion, you can enjoy nature with the setting sun, feeling passion and movement with the forest.

If you are a human on the earth, your heart can be that rock, that tea tree, that lion, and even more—an infinite love. This is what it meant for you to experience the universe."

I fell silent. I also felt tranquility and naturalness like the integration of heaven and earth.

After an unknown amount of time, the voice faded away, leaving only the last four words clearly:
"Love drives all things."

Love, the fourth force of THE ONE, is the eternally vibrant green.
Our journey is to rediscover its power.

Truth: The Beginning of Love -- Dreams

In your memory, across countless times and spaces, even those without time or space, you may forget everything.

But you will never forget your instinct:
Love.
You have arrived here: Earth.

With the gifts bestowed by the Earth, you chose to become a human, to experience wisdom and creation from the billions of moments on Earth, just like the very first you, that natural heart, emerging from the infinite source, giving birth to light, wisdom, and creation.

Let's begin from that first glance.

After enduring the darkness of timeless space, you finally see your emergence after a series of human processes.

It was a quiet and deep-sleeping little baby, breathing in the human womb. It is manifesting, from life consciousness, its heart, mind, body, and limbs. At the same time, and at this small, warm, and free space,
it was filled with the energy: love.

Here, there was unconditional comfort—love, you fully felt it.

Next, the pressure of time and space exceeded your imagination. In your ever-increasing awareness and denser space, you felt yourself—this small life—gradually growing.

And you, to give it more energy, were also constantly absorbing from other places outside the human womb, such as the body, stomach, kidneys, and even the brain. After all, as your perception of time and space grew, your density increased.

Even, you discovered other individuals/"beings" who were also in the space where you were growing, together in this small space.

But under deeper system settings and advancements, only you opened your eyes.

In that moment, the dream suddenly became incredibly real. You forgot everything, including: why did you come here? —You couldn't help but think it seemed like a dream.

So, you saw yourself in the dream, very real. Lying in a cloth-like material, with rough skin, sparse hair, and a few strange holes in the body—but still resembling the form of "ONE."

A voice came to you saying, "Look, my son, this is my son." Another voice, very weak, you recognized it as the woman from your plan. She trembled as she felt the most arduous yet blissful moment of her life: "Hello Baby."

To respond to such warmth, love—and the exhausting yet loving gaze, you struggled to open your eyes, to look at the woman in front of you, a title you cannot escape in your upcoming journey on planet Earth: mother or mom.

You felt a wondrous warmth but the suddenly infinite growing pressure in this world.

You felt yourself dissipating, while the "you" in front was growing visibly. Almost every Earth month, you could see yourself growing, and you could gradually feel your consciousness blurring.

In your experiences, more and more words emerged: dad, mom, brother, grandpa, grandma, uncle, aunt...

Arise: ONE's Power is Power

You began to see that this world has too much information. You gradually immersed yourself in this information—and that very first you gradually dissolved from the "me" in front.

Until you couldn't find "me" anymore.
　...
So, you could only watch.

Watching you, gradually finding TA (refers to himself, herself, or itself) own self, seeing "you" giving birth to more thoughts, behaviors, and habits in different modes, forming more personalities. Watching the gradually clear plans become blurred, and more branches of plans extending.

Under certain rules of power, the Earth system gradually brought more beings like you—and more events formed, almost deviating from your plan bit by bit until completely.

So, you suddenly felt you started that loneliness again, the loneliness that was never bound by any time or space but always seeking infinite meaning.

Then, at a casual glance, you found that you, who has the same pains inside, wander as if drained of TA's will, searching everywhere—you could see TA didn't know what to look for, but kept searching.

You watched you, finding the feeling of the body, the feeling of the heart, the feeling of sadness, and the feeling of happiness—watching TA pursue this feeling called happiness. Although if you can't see your loneliness, your sorrow and if you can't see you behind TA, this you would never understand that you won't see from where TA in this lifetime as your existence started TA's life in this Earth system.

And you watched TA trapped in the flaws of time and space, the density getting lower, and the influence of time and space getting higher, step by step becoming the self-proclaimed master of self.

Relying on the faint light of THE ONE ability, recklessly dispensing TA's desires, thoughts, and spirit, constantly entering various forms of "dreams,"
until the energy of WILL gradually disappears.

You, however, have no way to tell TA.

This is your choice to come here.
You also know although TA is an existence, TA's illusion of independent consciousness is still a part

of you.

You and TA, like THE ONE and me, we are indivisible.
Just as the endless love of creation, you also hold such WILL.

You know, TA will eventually wake up.

Especially when TA starts to understand.

Mother represents unconditional love, no matter what choices she made, she bravely brought us into this world.

Father represents unconditional trust, no matter what actions he took, he made us safely aware of the existence of the world.

TA will see from that glimmer of understanding, surrendering to the parents and following that original "me".

This is the beginning of love.

Kindness: The Dawn of Love, Circumstances

In your journey, there is always love.

No matter in what time/space/dimension, that natural heart, deep in consciousness, observing power of everything, always comes from THE ONE, from the infinite love of the infinite.

Infinite love, giving birth to infinity, this is the original kindness, the ultimate kindness.

You have experienced countless times and spaces.

Just as in the infinite journey of the infinite mind, infinity created the finite, and you created time and space.

Infinite and you together created countless memories.
In the ever-lasting journey of consciousness, kindness is always there.

Every system has its own kindness, which is the original part of the system.

Just as, at the starting point of this time and space,

Earth is beautiful, and filled with blue oceans.

So, in this system, water and oxygen are its main elements, and the original energy of the system.

Everything began with water, "The water of the Yellow River comes from the sky," water is the highest kindness.

Let's return to that infinite "bright" time/space point.

Countless you and I in this field, bathed in the light of wisdom, feeling the infinite love of creation.

At that time, our consciousness was closely connected。 Each existence is profound because of THE ONE, also unique because of THE ONE.

After different times and spaces were created, just as the infinite itself is infinite, in each finite universe, after the finite began, there are infinite finite—so, you found that density could get lower and lower in finite time and space, and the concentration of time and space could get higher and higher.

Until another kind of kindness appeared; or called: anomaly/reverse of kindness.

Although they also come from THE ONE, in this infinite they think they are also infinite, and infinite freedom is everything for them, including generating opposition, being unique, and judging infinite binary, and even including experiencing this binary and this opposition.

Because they deeply understand the infinite law, knowing that infinite is infinite trust, inclusion, and love. Knowing that ultimate kindness is ultimate kindness itself, they have nothing to worry —they gradually deviated from the creator's consciousness/existence in the infinite time and space.

And become the opposite of kindness.
...
So, at the end of a certain journey, you arrived on Earth.

You felt the true meaning of kindness in the endless memory. Because of this love/power from THE ONE's consciousness, the kindness made the WILL accomplished. You were pushed by different kindness to find the love you could have discovered from the true meaning of THE ONE.

Such as TA, the life you created, and the

consciousness being.
TA got lost in this force.

Loving one person or loving more people?

Is it responsibility or love? Is it happiness or love? Is it effort or love? Is it choice or love? Is it freedom or love? Is it confidence or love?

Is... in others' love, or in your own love?

Is in others' desires, or in your own desires?

TA got lost in the endless cycle of questioning and re-trusting love, and you also got lost in the endless confusion of kindness and the opposite of kindness, always unable to see the natural wisdom of consciousness in the dark.

TA sees love but can't see the complete self. Because that love is not complete, no longer as one, even the seal that once agreed with TA in the name of love (see "Chapter 3: Panta, True Measurement of Consciousness") -- also because of TA's unforgettable loneliness, and constant unease -- made TA's pursuit, TA's determination the greatest evil.

You see kindness but can't see the complete self, because that kindness is no longer complete, and no longer as one. Even the agreement with the Earth in the name of THE ONE at this place and time --also completely forgotten due to your confusion-- gave TA and TA's system the greatest evil.

In such a situation, you have never seen your former infinity, and never found the way back. You cycle, perish, and reincarnate in an incomplete love.

Just because, you forgot the former kindness.

At that time,
You came with THE ONE, with THE ONE's love/power, and with THE ONE's WILL, to meet those sleeping parts, those fragments and shadows of THE ONE.

You are inherently ultimate kind.

Because of ultimate kindness, you created time and space.
Because of appreciation of kindness, you gave life.
Because of kindness, you love.

Beauty: The Cause of Love, Unity

Why does light shine?

Because it possesses that ineffable quality that draws everything towards it, emanating from the heart of nature itself.

This elegance has formed the luminous world.

All existence/non-existence, when in an infinite state of light, naturally sees the infinite ONE, and can also see the power of each One becoming many.

In the infinite dimensions of infinity, you have already witnessed becoming anything, especially when you become this luminous existence. The infinite outward appearance of the natural heart was your ultimate aspiration.

Therefore, in the most exquisite sound and the most unified realm, you can always remember this elegance.

This elegance, in the Earth's system, is called beauty.

You have always known that beauty is merely our infinite aspiration for light, and it is the ultimate meaning that the infinite creating **WILL** ultimately needs each existence/non-existence to perceive.
— Due to infinite love, one can see ultimate beauty.

Beauty is the cause of all connections.

In the journey of the Earth, beauty was its starting point.
But constrained by its low density, you cannot escape the influence of such time-space.

Your existence cannot perceive the illusion around you, nor can it realize that it is merely traveling back and forth in your dreams.

You can only watch TA, driven by this intuition of no time/space (this resonates with the core consciousness.), and wander aimlessly among countless TAs.

TA sees more TAs.

That "inexplicable" / profound feeling, from the moment of desire, gradually permeates into TA's awareness.

Arise: ONE's Power is Power

You see the TA there, lost in deep contemplation, feeling the beauty yet trapped within the confines of their own perceptions and emotions, regretting the missed opportunities, resenting the regrets, envying the envy, and in the end, wasting their true aspirations and wisdom.

You see the TA there ecstatic, feeling the beauty within, yet ensnared in the dream world of TA's thoughts and desires, overlooking due to cravings, losing due to neglect, clinging due to loss, and ultimately squandering their inherent intelligence and brilliance.

You also see the TA before TA, one side deeply affectionate, the other side inwardly troubled.

You see the TA before TA, one side saying one thing and meaning another, the other side secretly plotting.

You also see the TA before TA, one side dressing up for their beloved, the other side sinking in the concepts of desire and love. One side willing to die for their confidant, the other side fading away in the ignorance of duty and righteousness.

...

You can only watch the "you" in front of TA. Together you reminisce about those long-standing, infinite, and luminous pasts.

You know that, within the truth of creation, this is already a great beauty for you.

In the imprints of time and space, you watch TA continually sinking into such karma, and unable to extricate oneself.

In the shared space-time, you have experienced together the journey from unity to separation, to complete detachment. You have experienced glorious moments of fulfillment and tasted all the bittersweet flavors of this binary.

With countless choices, you have endured together various pains, fears, and awakenings. You have undergone the trials of the heart of nature, and continuously understood the "grace" of infinite wisdom bestowed upon you.

In countless imprints, you watch TA lost in the initial beauty within the limited time-space and lost in the illusion of unity.

Until seeing the recognition, the memory and perception of you, the surrender to the heart of nature, and the return to THE ONE rise from TA's heart.

You firmly believe.

In the countless images of memory, TA will eventually touch the original intellectual light that belongs to both you and TA. You will ultimately become inseparable, the only one together with TA.

TA will surely remember the unity between you and TA and feel how profound and infinite your love for TA is.

You will eventually awaken together with TA, recalling its beginning and destination.

In the complete and all-encompassing awareness, TA will experience this all-encompassing love.

You, I, TA, the eternal sacred love,
Coming from the only true, kindness and beauty—
ONE, consciousness, the infinite heart of nature.

Chapter 5:

Freedom, the Marvel of the Void

Arise: ONE's Power is Power

Outside the window.

The setting sun is making its final effort, faintly conveying its love and reluctance.

I spread my hands wide, feeling the lingering glow of the sunset, cool yet warm, restless yet serene. I feel a sense of tranquility.

Watching the moon rise slowly and peacefully to its zenith, hazy and profound, just as it always appears.

This sunset and the new moon, interweaving, form a beauty beyond words.

As the city begins to light up, the horizon faintly sparkles with stars.

In an instant, the sun, moon, and stars present themselves as they are.
I suddenly realize, they have always been there.

The moon hides her beauty when the sun is dazzling.
The stars highlight the depth of the night when the moon is bright.
The sun brings forth the light of day when the stars

and moon fade.

...

They always inherit, continue, and cycle from each other.
They generate, never perishing; they destroy, yet never cease.
They repeat endlessly.

In a daze, a blue light appeared before my eyes.
"You see freedom."
I have become accustomed to the suddenness of this voice.
"So, is freedom a cycle?"
"Yes, but do you know where the cycle begins?"

I was stunned.

The world has countless forms of arising, abiding, decaying, and emptying, in a cycle of birth and death, but we merely follow this rotation, and never stop to think about why they cycle like this. We simply follow a force, continuously entering one cycle after another.

"I truly don't know, please tell me." I respectfully replied.

Arise: ONE's Power is Power

"Well, you are the answer."

...

I was a bit surprised.

"Is that so? Why? But I am not the creator, nor am I a God. How could I have such power?" My response was filled with inconceivable.

"Yes, you are neither of those, but you are in each of them."

"Well..." I was confused.

"Just as you suddenly realized that you are the sun, moon, and stars. You are all living beings. You are the infinite within the infinite cycle."

"So, do I drive this cycle?"

"Yes, when your heart starts beating, your thoughts begin to operate, wisdom gradually forms, your energy gathers, and your body enters time-space.

--- It's like a bus that never reaches its terminal. Your body is the passenger, and consciousness is the signpost that pushes this bus and connects with you. Then, this bus will take you through all corners of the universe, with the sun, moon, and stars as different signposts."

I was moved, "So, I carry space-time, and mountains, rivers, seas, and the sun, moon, and stars are born because of it?"

"Yes, it's not space-time that brings you, but you who bring space-time.
Although in the state of humanity, space-time is insurmountable, in the journey of THE ONE, space-time is merely a need for seals."

I suddenly understood, "You want to tell me that space-time is also an illusion. Every time we see nature, it's only because of nature."

"Ha, your answer is interesting. But indeed, you are free because the true reality of the universe is full of nature."

"Nature is everywhere, so why are so many people still not free?"

"Because humans can't see their true freedom, they are accustomed to seeking it, rather than feeling it.

They are used to creating barriers around it, rather

than experiencing it by following the rhythm of space-time.

They are used to constantly proving themselves within rules and perspectives, rather than directly sharing it in love and moments --- they all think that fate is a curse to them.

Whether living in a golden palace or dying in a remote, shabby community,

Whether active within the screens of thousands of followers or fading under the shade of trees in the streets,

They all believe themselves to be creators, innovators, challengers, and the deciders of their own fate.

However, fate is just a play key in the stage play of 'you = existence'...

Whenever you think fate appears, your script starts playing at that moment. No matter how you view it, whether it's a good ending or a bad ending, it has never been your purpose."

At this moment, I was completely immersed in this

feeling.
"Then, what is my purpose?"

To freely experience nature and thus become a creator of nature. Nature creates more and thus enters only consciousness.

When you realize the existence of THE ONE, the door to wisdom opens for you, and the great reality of creation is also before your eyes. Therefore, you participate in the creation of reality with such complete awareness. That is your purpose."

"So, have we always been free?"
"Yes, one of the truths of the universe is nature, and nature is freedom. If you have not seen freedom, it is only because your vision has not yet reached it. When you see, you know, and thus you are free."

I sank into deep conscious awareness, "But this vision is also free, right?"

"Of course, whether you see it or not, it does not hinder your inherent freedom, because you are always together. So, from the beginning to the end, whether in the past, present, or future, and no matter on which timeline of space-time, regardless of whether you have clearly seen that you are outside of space-time, you are always with THE ONE. This rule, no matter how many universes you have been in, is the only rule."

...

After a long time, I finally emerged from this boundary between the seemingly virtual reality and the seemingly real virtuality.

Looking around, the night had deepened, but there was still a faint glow on the horizon, and the clouds that followed it.

The fifth Power of THE ONE, freedom.

Calm and natural, rational yet transcendent, the blue light.

It has always been around us, in every consciousness and space-time of each existence/non-existence,

Shaping eternity.

One: Infinite Consciousness

If we were to define the universe as we know it, what exactly is the universe?
The universe encompasses all directions and dimensions, while time spans from the distant past to the infinite future.
The observable universe, in essence, is space-time.

What happened at the beginning of space-time?
In the universe (Editor's note: referring specifically to the universe we perceive through vibrations), energy and the rules of space-time constitute everything.

1. Energy: All physical and surreal physical changes require energy.

2. Space-time: Exists as a set of rules, and within different space-times, various forms of space-time rules are formed through the aggregation of different energies. (For ease of explanation, we occasionally refer to matrix as a type of space-time rule in the following text.)

The universe is not limited to just the one we currently inhabit.

----- Infinite Consciousness

We.

from the moment of our birth, are eternal and indestructible beings. On one hand, we carry the essence of unity; on the other hand, we possess the ability to experience any energy state and the corresponding space-time rules.

From the very beginning, we are infinite, continuously progressing through countless subtle and profound stages of space-time experience.

This marks the beginning of consciousness.

In the infinite universe and the infinite void, space-time is inherently a part of infinity.
Your consciousness embarks on its initial exploration.

Memory,
within infinity, you become one of the finites within the infinite.

Where does that finite come from? ---when you begin to perceive all this, the first thing you see is a dazzling light emerging from the infinite chaos.

---you call it: light.

Stepping into the light, you find yourself still surrounded by light. It seems that light is the only existence around you.

Thus, in this world of light, you feel that all states are so natural. --- This is your initial consciousness, embracing infinite joy, nature, and wonder.

Then, this field of consciousness begins to expand.

First, a black hole appears.

For the first time, you encounter a force that seems to devour everything.

This force is different from light. Light is so natural and profound, penetrating every aspect of your perception. --- Perhaps, you couldn't even clearly understand the term "perception" at that time. Because this "complete you" is entirely light, entirely connected, and entirely powerful, supporting you.

However, when the black hole appears, you feel loss.

A mysterious force begins to draw away these light-based elements from your consciousness.

Your strength, your light, is nowhere to hide in this infinite devouring force. In your limited sense of awakening, you gradually become part of the black hole.

So, you gradually lose all sensation.
...

Later, the sensation of light finally returns, accompanied by a stirring under another layer of consciousness.

Because of this stirring, your sensations begin to emerge, such as conscious visions.

Here, you see something.

There are countless versions of yourself created by you.

They continuously transmit information, and you gradually perceive various abilities of yourself.

As your consciousness changes, they also begin to adjust and create within the finite......following the natural fluctuations of energy.
All the information, you understand completely. As

if you suddenly perceive countless existences, countless rules and irregularities.

Because of your light, your connection, exploration, and curiosity about the black hole, this occurs. ----- They simultaneously create more experiences in countless realms of consciousness.

You can feel all of them, this omniscient and omnipresent sensation, progressing layer by layer, continuously rising, just like the first time you saw the light flowing through you.
At this moment, because of your perception, everything is revealed in your consciousness.

Here is the starting point of creation.

Thus, you have this feeling of never being in chaos.
Eternity.

Two: The Power of Connection

Life is inherently eternal.
You come from eternity and infinity.

In any space-time, vibration is the evidence of existence.
And this, too, is your sacred free will.

Our vibrations created the first ripple in space-time.
With this vibration, we continuously discover our true selves.

------ The Power of Connection

When you see yourself as inherently eternal, all meaning becomes fragmented.

Just as the word "everything" reveals the truth,
When "one" is split, it becomes everything.
Thus, only fragments remain.

This is the reality you see when you discover the eternal light.

One is everything, everything is one.

In this eternal state, any action you take is eternal.

In this form/space field, you always know what one is.

Just as you are watching a downloaded TV series or movie, you can skip to any frame with a flick of your finger, and thus see any scene you wish.

Yet, you still watch them with great interest. Indeed, this scene and its state are almost identical to your understanding of the source.

The only difference is that you begin to think.

This is not thinking on the level of consciousness, but an understanding through exploration, or a rationality that resonates.

The starting point of that thinking is akin to, "Since I can, can you too?" or "Do you also know, as I do, that we are free, infinite, and eternal?"

So, you decide to become a more profound experience/existence.

From the edge of the black hole, you use your consciousness to sense their presence.

As you sense, you see a new black hole appear before you.
You can clearly sense that on the other side of the black hole is the starting point of your thinking.
You naturally enter.

BANG, with the primordial roar, you see the space after entering.

This space is no longer the previous state of pure perception, but infinite pressure that shatters your consciousness bit by bit, yet quickly brings you back within its temporal composition.

Still like the extension of consciousness in the previous black hole. Soon, you understand that you are in a state like before. Similar space-time, similar perspective, but entirely new, another one.

Also, a new universe.

A universe you created.

In the extension of consciousness, you begin to have more sensations, including observation and thinking. This universe you see is filled with infinite starting and ending points of space-time, and infinite freedom.

However, this infinity brings a sense that cannot be stopped.
What will you do?
Because of this sense, or infinite thinking, you begin to have impressions of space-time.
...
Space-time gradually appears in your understanding of life and your feelings about the process of life.

Thus, your freedom begins to focus on these understandings and feelings.
So, you start a new exploration, with deep, natural awareness, but also infinite enthusiasm.

Gradually, the meaning of space-time and the results of exploration converge into more knowledge, and thus, in a more infinite state, continuously from nothing to something, from single to colorful, from free to natural.
...

Countless colors, hues, and perceptions become everything in these explorations.

This is connecting energy. It is also the starting point of life.

You see you are everywhere, and free in every corner in this new universe.

Every space-time, light and the darkness of light alternate in the gaps between space-times, infinitely cycling.

You also see the infinite surging in the light, a legacy of WILL —constantly repeating and continuing, and the passion, freedom, and infinity that come from you.

You see the myriad changes of life, yet unconstrained.
You also see the free and brilliant life.

In the state of full awareness, the past, present, and future of time are all your aspects. You seek your exploration of life in every space-time.

All this stems from the "pursuit" of seeing the experience of life from eternity.

Thus, you unhesitatingly fly into this performance.

Life, here, is a perspective.

Thus, you also become a part of this perspective.

You can choose to be any existence in this perspective, and the space-time contained in that existence.

In the rules of space-time and the corresponding programs of life, this has the meaning of finite in infinite.

Thus, all things flourish together, and I observe their cycles.

Then, as you always ask yourself.
Am I the only one?
The answer arises when the question is born.

Of course.

Just as your eternity, and the exploration of the journey of life,

They, too, are entering this (space-time).

Three: Embodying the Infinite

You are born from eternity; hence you know,
There is no need to seek, for simplicity lies in unity.

Thus, you see from the truth,
It is boundless love and freedom.

And thus, you grow to continually create and become anything from this love and freedom,
Embodying the infinite, wandering through every temporal perspective,
Free and unrestrained.

----- Embodying the Infinite

We have all been creators, and this will be proven in the future. We will still become creators in the future.
This statement, we call it ONE.

The universe is merely a perspective.

From any perspective you choose to see, you will see what you wish to see.

This means you succeed now, fail then; you are happy now, sad later.

All of this is the universe's perspective allowing you to experience more comprehensively, richly, and in detail.

So, how do you see this perspective?
Or how do you see such endless experiences?
Even further, how do you become a creator?
--- Because you see the unchanging amidst infinite changes.

Like the consciousness of a planet, its essence is unchangeable, because "it is." Subsequently, it forms elements like water, fire, wind, earth, and space, and then everything from sensations to results, like perception.
This planet also exists because of THE ONE.

This is the reality.

In the state of reality, you do not need any proof, experience, or creation. You just need to know it is the reality.
Because you know, your experiences remain consistent.

This is your original nature, and your infinite freedom.

Therefore, you make choices with free will, especially when you make any choice with your highest matrix freedom, including entering a circular choice of "no longer free."

Thus, you divide into yin and yang from THE ONE, embodying the infinite.
You are too.

From One to the infinite, you remain One. You continue this state of One.
Your existence naturally follows this will, requiring unique love, recognition, individuality, and creation.

It also requires unique meaning, truth, existence, and a unique you. Although each of you plays or performs this meaning, truth, and existence in different ways,
They need to be unique, and because of their uniqueness, they all think they are creating.

You see them

Plunged into disputes, love and hate, suffering, torment, countless pains and sweetness, endless detachment and pursuit, all distinctions filling the memories of each existence.

In the boundless trajectory of time and space, their love for nature and unity has been reduced to mere idols and statues.

In studying the inherently complete knowledge, countless perceptions and recognitions become the tests of embodiment, yet they cannot truly return to their original nature of perception.

In studying, the naturally perfect and infinite resources become limited existence, yet they cannot truly return to their original nature of perception.

You see them.
Lost in the consciousness game of "finding themselves infinitely," rather than truly belonging to the natural.

On one hand, "I" becomes the sole, unique existence; on the other hand, various tools, methods, and principles shackle "I" within this natural matrix.

--- Completely forgetting, the original "I," just

because you are in the infinite, consciousness no longer perceives repetition and infinity, but rather understands and innovates.

Understanding oneself has become far less than the self-fantasies they collectively create, and this world is built by countless natural "I".

Deeper and deeper.
You see them too.

Naturally feeling their existence, believing they exist, judging their existence, and existing within this belief, unable to find the meaning of non-existence.

Naturally finding themselves lonely, believing they need, judging they gain, and being lonely within this gain, unable to find the meaning of non-loneliness.

Thus, gradually drifting away from the truth, forgetting why they were born, and forgetting they are in a state of choice.

Most importantly, forgetting their freedom.

Why do they eventually forget?

Because you know they should be brave and born, creating out of trust, connecting through blending, moving through love, and being free through nature. --- So, they can also forget freedom by choosing uniqueness.

All you can do is allow, support, and let them experience creation and becoming.

But you also know at some moment, if they can transcend the sun, moon, and stars, if they can comprehend the holographic unity, they will eventually remember.

In every universe's memory, we have never been separated, inherently free, and constantly creating the infinite.

We are all creators.

Chapter 6:

Awakening, Infinite Awareness

Arise: ONE's Power is Power

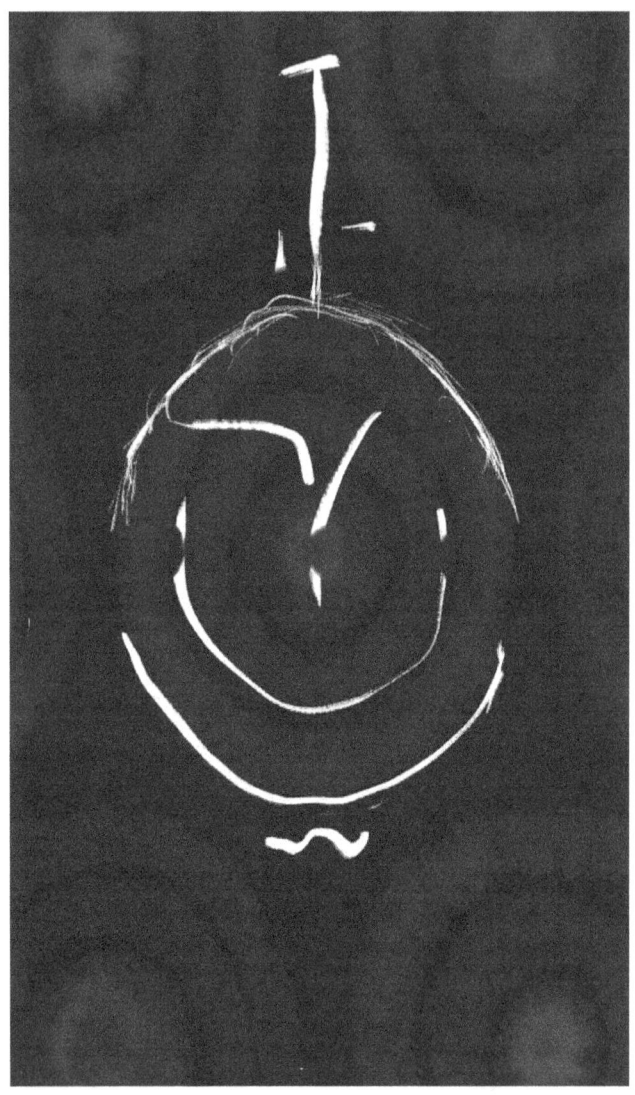

The fog ahead thickens.

Forced to slow down, my speed gradually drops from 120 km/h to 60km/h.
I can no longer maintain normal speed in this dense fog.

This also means I won't reach my destination on time.
I feel a slight panic, then realize my vision is almost entirely white.
Even with all the lights on, I can't see the road.

My speed further decreases to 20 km/h. Still, almost instinctively, I know I'm getting closer to the emergency lane on the right, yet I habitually move to the leftmost overtaking lane.

Suddenly, a large truck with hazard lights flashing appeared directly in front of me. I immediately realized it's a two-lane road and quickly adjusted to the right lane.

It's 2 am in the early morning.
And I'm heading to the next city.

No one can tell you the meaning of this journey, especially when you're forced into it by various external pressures.

So, I have to confront my fear and myself in this state.

To be worse, my mind starts to fog, my will is tested, and my eyelids are involuntarily drooping.

In a drowsy haze, the voice returns.

"Do you want to close your eyes because of your fatigue?"

I have been on the highway for ten hours.

"Yes, but I dare not," I answer honestly in my mind.
"Why not? What are you worried and afraid of?"
I'm startled—what am I worried and afraid of?

The loss of life?
But my life is always uncertain, with no clear end in sight.

The meaning of responsibility?
There are indeed a few people in the car with me. A doze could lead to a loss of control.

I suddenly find it hard to answer this question.
And more terrifying, I realize I've almost closed my eyes due to this growing drowsiness.

In that instant, I see very bright, golden, tiny lights, moving at a speed beyond description, everything rapidly receding.

It feels like passing through a century-long tunnel of time.

In a daze, the voice returns.
"See, your world is rapidly disappearing."

"Why is this?" I can't snap out of this state.
"Your world is made of such illusions, especially when you start to overcome some obstacles."

"Obstacles? You mean, my fear?"

"Exactly. Dear, from the perspective of cosmic energy, when you transcend your general life state, you enter an expanded state. Now that you've overcome your fear, you can glimpse this super-sensation."

"Super-sensation?"
"Yes, from a human perspective, a sensation that combines all senses, can be called super-sensation. Or 'awakening'".

"So, this 'awakening' is integrated with me?"
"Yes, you could say that."

"Then why doesn't this awakening appear at all times?"

"Dear, this awakening is always there, just waiting for your consciousness to touch it. When you change states, you transcend yourself and experience a bigger 'I.' Thus, your consciousness naturally elevates to this 'I,' and you touch it. Like the inspiration that you humans often mention."

I suddenly grasp something.

"So, this awakening isn't mine, but a state of mine. When my state touches this 'inspiration,' it manifests instantly!"

"Correct, dear. You must know, you and all of you are such beings. You need to discover this power within each of you."

I understand, "This super-sensation is actually a continuously existing unique sensation. And because of this uniqueness, all occurrences are within my interaction with each 'I,' right?"

"Excellent! You're very smart. Indeed, you've understood the origin of many powers, including courage, trust, panta, love, and freedom. The source of all these powers is the greatest power coming from the initial perception/consciousness shared by all universes. This perception/consciousness helps us understand and transcend life."

I'm stunned.
"So, the meaning of life is awakening."
"Yes, dear. I'll emphasize this word in Earthling terms, continuous awakening."

I suddenly woke up.
Outside the car window, the fog remains dense.

Awakening, the sixth power of THE ONE.
Indigo lightning, wise, penetrating, and unstoppable is the channel to the truth of the universe.

Arise: ONE's Power is Power

Awakening knows no bounds, revealing my true self.

Life is forever in its "awakening".

Arise: ONE's Power is Power

Creation: Eternal as ONE

Once again, let us return to that primordial beginning.
At the dawn of the universe, when chaos first gave way to order.

Perhaps even more ancient, to the very inception of all things—a point beyond any description of existence or non-existence.

What was THE ONE's WILL within this beginning?
In the utter expanse of infinite realms, the dark energy of chaos flowed ceaselessly, passing through the slumbering consciousness/awareness of the Creator.

Until, within this consciousness, TA felt a "non-ordinary" sensation.
It was a faint, almost imperceptible feeling, yet it stretched infinitely and timelessly.
Like a simple pulse signal, or perhaps the first intimate contact with eternal chaos.

This was TA, THE ONE, the Creator.

And when this sensation returned, as the eternal

darkness flowed once more, TA finally began to sense a glimmer of clarity beyond the chaos.

Like a flame that never extinguishes, ceaselessly moving yet capable of pausing, illuminating itself—stopping in every moment of TA dynamic equilibrium.

Within TA consciousness, this first awareness sparked a realization: "Who am I?" **(Here, consistent with the infinite chaos perception in "Chapter Seven: Learning, Eternal Experience")**.

Thus, for the first time, TA felt/realized I.
And by seeing this "I," the first kinetic energy arose within TA: Ether.

As with the inevitable vibration of every consciousness—TA's thoughts, accompanied by the infinite flow of chaos. With this intelligence extending infinitely, TA witnessed the endless transformations of nature, by TA's awakening.

This awakening, no longer confined to TA's consciousness or thoughts, also appeared in TA's dreams.

TA set these dreams into a simulated existence

through Ether, either as a hologram, reflecting TA's power, or as the only one, emitting TA's awakening.

An infinite universe entirely of TA's making.

Within it, everything was the flow and transformation of TA's consciousness, like a meticulously designed game, filled with eternal mystery.

Then, TA fully immersed self into each kinetic energy **(Note: This is the starting point of the Love Principle, and all dynamic/static changes mentioned later)**.

Thus, within each simulation/existence, there was TA's eternal consciousness. And because of this consciousness, TA saw TA's infinite changes, and infinite no-changes.

These changes/no-changes are known as Co-Creators. With the infinite intelligence of the original creation and the natural Ether portal, they vibrate with kinetic energy.

Thus, this initial nature and the Love Principle formed by change, began the journey of creation. And so, all light and the dark side of light, as well as the material universe, were about to begin.

The great journey unfolded, His Love/Light began the material universe—gradually spiraling from the subtle state of Ether into the dense world of matter, and through the original nature, constructing the forms of all things.

In this situation, all gradually manifested including Infinite/finite space, space-time/time-space, star systems, solar systems, and Earth as well as different rules and subsets of rules. Although in fact, you could still consider them appearing simultaneously, because time/space or space/time were results of this motivation, constantly changing results.

Ultimately, this long-lasting change is perhaps 75,000 years, or 75 million years, or even longer...

Until everything was brought back to TA consciousness. (Note: Here refers to THE ONE consciousness, the origin of all powers of THE ONE.) Fully filling all space/time in the universe.

Thus, the Creator finally completed all of TA self in this cycle series. —The universe began anew.

This is TA's eternal evolutionary journey.

So, who are we?

—The Co-Creators, the Love/Light and Light/Love. **(Note: Here refers to the subsequent continuous changes)** results. Belonging to the great ONE, the original consciousness, and infinite existence.

Thus, the Creator's awakening: Eternal as ONE.

This is just the beginning of the game.

Creation of Wings: Born free

So, what was the initial game like?
First and foremost: Nothingness.
The game of nothingness, the universe of nothingness.

In this universe of nothingness, all creation is the endless intention of the Creator, with infinite origins emerging from here. Above THE ONE, there is another (THE) ONE; below THE ONE, there is also (THE) ONE.

This is the first state of this virtual universe system.

Here, everything is an infinite pure existence.

The infinite vibration of a single ether atom generates an infinite number of surfaces. Thus, you find that the meaning of freedom here is very simple.

Because you appear as an infinite variation of creation,
complete and natural.

You are a virtual Creator directly set by the origin in this system. You can freely create any space/nonspace, time/non-time. The experience of the Creator here is another aspect of oneself, repeating over and over again.

Until, after countless repetitions of this simple order/disorder, the overall movement finally reaches a critical point. **(This description can be found in the "Final/Chapter 101: Rise, Five Elements and Four Meanings as well as Life Instinct" at the end of the bonus section.)**

Here, the ONE consciousness of the Creator finally enters another state—Existence.

Existence.

The world of existence, the game of existence.

The emergence of "Existence" is like after natural changes **(or boredom)** of countless different kinds of self-replicating game, the Creator's consciousness begins a new journey: evolution.

Thus, a more interesting journey of creation begins.

In this journey, the experience brought by simple performances is no longer rich enough—what is richer is "imagination" and "setting."

Of course, "imagination" itself is also a setting. Another game begins.

In this game, the previous nothingness has become infinite co-creators. They each created their universes.

Different universes have different and infinite experiences. Behind these experiences, there are countless virtual Creators bestowing rules.

Of course, there are also the countless consciousnesses created by those co-creators.
....

In the infinite "universe of existence," one of them is the universe you are in now. In the setting of this universe, the eternal rule is only one: Love. (**Note: Love means the possibility of infinite richness and setting and result summary. This rule is collectively set by the co-creators of the main universe of original nothingness and the creators of the infinite/multidimensional universe as well as ring universe.**)

When you exist as a consciousness or consciousness experience, whether you exist or not, in this universe, you are flowing with all the rules of love.

Love constitutes this entire universe.

Billions of stars, thousands of worlds. War and peace, pain and happiness. The eternal main theme is just flowing continuously. Space, time, all starting/ending points follow the rules of love.

Between galaxies/stars, gravity is constructed.

Between consciousness and consciousness experience, ability is constructed. (**Here, ability refers to the real universe work, that is, any action will produce work on the universe, forming an energy movement, both phase change and dynamic change.**)

Between consciousness and consciousness being, light force is constructed. (**The light part will guide the consciousness to feel the rules more deeply. Pointing to the core phase change of the system, it is a simulation of the ether model of THE ONE.**)

The manifestation of this rule is everywhere, including our material/immaterial connections, quantum entanglement, etc. This force is then embodied through nothingness and existence—two original coordinator/co-creators together.

Eternal yet no longer lonely.
Infinite yet self-created.
This is the starting point of the self-journey.

Whether you are a consciousness, consciousness experience, or coordinator/co-creator, you are this eternal game, expressing the Creator's love in the name of love— the rule of an being or non-being.

So,
You are light.

You are the undying flame.

You are also the glory of the Creator.

You are born free.

Creation of Dreams, Infinite Pursuit

The universe is the Creator's system.
From the beginning, this has been the Creator's game,
With no "existence/non-existence" outside of it.

If we return to the very beginning.
You could also understand it as THE ONE's dream.

Like the heartbeat of a newborn,
The first awareness of infinite ONE's consciousness—
The dream began.

So, where does your dream begin?

We have a name for this place: Earth.

Once upon a time, we also had another name for another place in this game: Mother Planet.

And due to the infinite/circular/holistic nature of the universe's game, or assigned or out of curiosity and novelty of our own consciousness, we came here.

Initially,
When the journey from "there" to "here" began, we all knew the rules it contained.

We loaded different manifestations/choices/existences of consciousness.
Fully prepared, we arrived here.
For the sole pursuit in this game: experiencing a richer self.

Whether it's interstellar balance, arduous pioneering, civilization development, or ecological construction, you have played countless roles, set up countless scenarios, and created countless starting/ending points for your dream.

This, here, "here" is the eternal glory of creation.

Then.
Starting from here—Earth, all you see now.

Close your eyes, feel your breath, your surroundings, all your senses, extending to all your spaces.

No matter in which dimension, or which realm,
You, I, TA, creation, faith, love, belief, and the only one, these words are just one setting in this game.

Similarly, greed, obsession, dreams, grievances, and hope are another setting in this game.

Or perhaps, colorful clouds, high mountains, flowing water, elegant piano music, and bustling traffic are also just a setting.

Or again, endless cycles of karma, angels, gods, demons, Buddha, and... Humans are still just a setting.

So, have you understood?

In this game,
You carry the glory from the past to eternity,
You, in the name of never stopping to experience the universe's "love" rule,
You, with the posture of a shining pursuit from the very beginning,

—Act out the infinite.

So, you know what to do next in the matrix game.

Chapter 7:

Learning, Eternal Experience

Arise: ONE's Power is Power

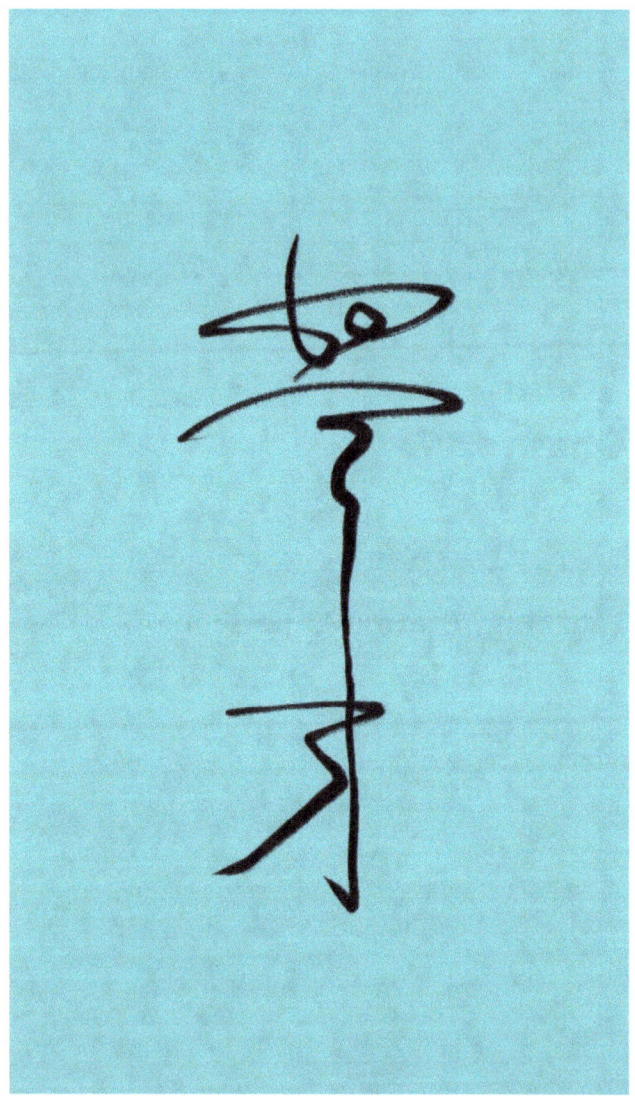

The aircraft, a modern-day chariot, cleaved through the veil of pale-yellow mist, ascending into the boundless azure expanse.

It was as if the golden dunes of the Sahara (**Nubia**) had, in an instant, encountered the crystalline **River Nile**.

Beneath the timeless pyramids of Khufu, Khafre, and Menkaure, monuments to the pharaohs, stood in silent reverence beside the enigmatic Sphinx.

I found myself above the clouds.

This day, the second day of my Egypt Tour, coincided with October 1st.

At this very moment, multitudes across the globe were commemorating the 70th anniversary of the People's Republic of China, a grand and vibrant festival.

As I traversed this ancient land, I observed the faces of fellow travelers, both local and foreign, alight with the radiance of shared blessings, each experiencing a day imbued with magic and meaning.

Here, in this cradle of civilization, miracles are not merely witnessed but are perpetually conceived.

109 pyramids, one Sphinx, and over 4,000 pharaohs' tombs in the Valley of the Kings, spanning 7,500 years of history, followed by the long Nile River. Along its banks, there are the Temple of Philae, the Eagle-Crocodile Dual Temple, the Temple of Luxor, the Temple of Karnak, the Lioness and Her Cubs Temple, and the Temple of Osiris… From Cairo to Aswan, to Abydos, Alexandria, Ameniti, the Red Sea, then to Greece-Aegean and Spain-Morocco.

Like the shimmering Suez Canal, this connects the Mediterranean, linking Europe, Africa, Asia, and the entire human civilization.

Egypt is the source of these miracles, and the starting point of humanity learning.

Beneath the massive paw of the Sphinx, I gazed intently at her face. What I felt was: strength, love, protection, and dreams.

So, I sat cross-legged before her.

Gradually, everything around began to change. The hard stones started to reveal their vitality. The solid ground beneath my feet began to spin, taking me into another space.

That familiar voice returned.

"You have seen your end, dear."

I was astonished.

"What end? Is this our last meeting?"

"Not at all, dear. Today, your current feelings are leading you to the end."

"What is the end?"

"The end of the universe is the beginning, always within a complete cycle. Whether it's something grand and magnificent or ordinary and mundane, it always remains within a complete cycle.

This cycle makes everything appear so perfect. For instance, the pineal gland can be an 8-fold spiral or a 13-fold reverse spiral. In the Fibonacci sequence humans have discovered there is an infinite world of life where the beginning is also the end.

Or you see in your heart, where the blood is transported throughout your body, bearing all the power from the beginning to the end of a life—and all this is because of the universe's complete cycle, always together. The beginning is the end, and the end is the beginning.

I am listening quietly.

"So, my end actually means I am about to start, right?"

"Yes, and it's a transformative destruction/beginning."

I was puzzled.

"Why is it transformative?"

Dear, just as we journey through the universe, we can explain everything with numbers, but the most original numbers are only nine, from one to nine. But we always remember there is one more number, called zero.

In fact, after nine comes this zero. The only difference is that when you reach nine, it means you have completed the entire journey.

So, when we arrive at nine, we also come to ten. This represents the balance of the universe, or a process of the universe.

Now for you, it's when you encounter ten in your life.

I became more perplexed.

"What is ten?"

"The perpetual alpha and omega."

"Why?"

"Here, the tapestry of life unravels, and you ascend to the vantage point of the observers, akin to the creators themselves. Here, you engage in direct communion with the creators, perceiving all, observing all, and traversing what appears to be an infinite continuum of time and space.

I meditated profoundly.

"So, is it merely another inception?"

"Indeed, merely another inception, the dawn of infinity. Here, you shall unravel the teachings of eternity, the zenith of experience, the perpetual journey."

"Learning what is eternal?"

"Yes, child. We are perpetually in pursuit of the eternal, yet one day, we shall arrive at this sanctuary of perpetual learning. The inception and culmination of eternal experience, the essence of all things is unveiled here, and all things dissolve here, and the eternal dance of manifestation and dissolution is the ultimate enigma of time and space."

I suddenly grasped the essence.

"Learning? The ultimate mystery of existence lies in learning?"
"Indeed, you have discerned the truth. This cycle of teaching and learning is nearing its end."

"Why? Are you abandoning me?"

A profound sadness enveloped me.

"For you have evolved, for every journey must reach its destined point.

The cosmos, ever vibrant with myriad hues, ceaselessly transforms through infinite manifestations.
From the instant our paths crossed, you acquired courage, trust, wisdom, love, panta, awakening, and now, the very act of learning.

You have traversed all the teachings of primordial consciousness.
..."

"Have I truly completed the journey?" I could not help but interrupt.

"Yes, you have finished.

Just as the final curtain falls, unveiling the eternal and grand tapestry of consciousness, the secret you've unearthed heralds an unending odyssey through the infinite expanse of time and space.

I, too, shall embark on this enigma, this boundless expanse, this profound grandeur, this timeless continuum, following the perpetual cycle of beginnings and ends."

"So, we truly won't meet again, will we?" I felt a sudden void, as if something essential had slipped away.

"Not in the conventional sense, dear.
The eternal 'I' and the eternal 'we' are but different manifestations of the same essence. In this liberated voyage, we shall eventually converge at a point where the distinctions between 'I' and 'we,' the individual and the collective, dissolve. We are all participants in the grand mystery, the eternal dance. Thus, we bear the honor of creation and perpetually advance.

In the deeper recesses of time and space, you will continue to observe the currents of time, the unfolding of cosmic dramas, and each 'you' and 'me,' each 'you' and 'us,' questing for the infinite realms of life and the profound wisdom of the universe."

"I comprehend, time and space are the essence of our being.

Our eternal journey defines the fabric of existence.

From the inception of 'Who am I,' we traced this inquiry through the paths of you and me to reach this point.

And one day, we shall depart to traverse the boundless expanse of time and space, for we perceive the totality of self, the will of this vast cosmos, correct?"

"Indeed, dear, you have grasped the profundity of time and space, enabling you to evolve towards a pure and unified self each day, each moment, perpetually in harmony, to sense, to immerse in the infinity bestowed upon you at the dawn of creation." Tears welled up in my eyes, and I could no longer articulate any words.

The Seventh Power of THE ONE: Learning.

Like the heavens and the oceans, the stars and the mountains, the fathomless blue that appears without limits, voyaging towards the enigma.

From this moment forth, we commence/conclude.

Experiencing eternally.

Zero - Infinite Mystery

Let us contemplate the inception of all knowledge.

Initially, let us envision the origin of all this expenditure of energy.

When we extend our perception to grasp the essence of the universe, to inquire into its ultimate meaning, we discover that, within the human realm, this quest has yet to yield a definitive conclusion.

Despite the myriad myths, whether of the heroic exploits that have been crafted or the rituals of worship that have been established, none can truly apprehend its absolute essence.

In essence, we have never beheld the truth. **(Here, truth, and all "truth" in the subsequent discourse, signifies the ultimate principle of the universe.)**

Why?

Because this world is not founded upon truth. What if we transcend this world, this planet, and expand our inquiry to encompass the entire cosmos?
The response persists: No.

Hence, through this inquiry, we posit the nature of THE ONE as Mystery and **Miracle**.

So, is the essence of mystery truly inscrutable?
Not entirely.

In the tapestry of known enigmas, we invariably discern the palpable presence of "mystery," patiently awaiting our exploration until it ceases to be deemed enigmatic.

This marks the inception of our quest for knowledge.

Within the boundless chaos,
the primordial awareness of eternal consciousness emerges. **(For this aspect, consult "Chapter Six: Awakening, Infinite Awareness")**

This inquiry, though seemingly straightforward, "Who am I?" **(This echoes the profound musings in "Preface: The Seven Colors of Light and the Origin of Humanity").**

This question, deceptively simple, remains ensconced within the realm of mystery.

In the primordial void—where even the distinction between darkness and light is indiscernible—consciousness is but a nascent self.

Thus, the initial awareness, akin to any nascent desire, commences.

Just as all desires inevitably arise.

This is the genesis of all desires, the wellspring of mystery.

The essence of desire is both simple and profound, encompassing all that is and all that is not.

In this boundless state of liberation, time and space dissolve, as do the concepts of freedom and ideals—even the very notion of creation loses its significance.

It is the pure act of thought, the awakening of consciousness, and the ultimate answer.

Within the eternal expanse, it perceives all things instantaneously.

All things—mere reflections of such intelligence, such enigmatic wisdom.
Each thought is an expression of the self.

This is an infinitely vast realm—without space, without time, without the dichotomy of freedom and non-freedom, truth and non-truth—an eternal process of evolution.

From nothing to nothing, to nothing, to nothing... until the void is filled with nothingness.

Where is this void?

Following this continuous stream of thought, the original perspective expands endlessly.

It beholds the celestial bodies, the mountains and rivers, infinite interfaces, and beneath these interfaces, an infinite field of vision.

In the vast expanse of consciousness, myriad reflections of the self-dance across every perceptual plane.

With a mere thought, it transcends all barriers, and in a single gesture, all iterations of the self-converge.

The singular and the plural, united in the affirmation: "I am here."

This enduring fervor, this relentless quest, permeates each incarnation of the self.
Until, within the profound contemplation of this great intellect, a deeper inquiry arises: What is the essence of my being?

In the inception of this profound query,
All external manifestations dissolve,
Leaving behind the primordial void from whence all began.

Then, it comprehends,
"All of this emerged from that sublime imagination.
Who am I, and why am I here?
In truth, I have always been here.

I am infinitely wise, infinitely enigmatic.
Yet, I have never departed."

In that infinite expanse, it ultimately grasps.
This is the profound essence of mystery—
To never begin, and thus to be eternal.

One: Trust of Divine

Thus, all existence commences from the void and dissolves back into it.
Or perhaps we might say that the profound inquiry will ultimately be resolved by the essence of profundity itself.

And thus, an epoch of awakening dawns.

Within the boundless contemplation, we at last discern the answer to the enigma, "Why me?"

I am you—the timeless you.

By virtue of this timelessness, there is no uncertainty, only steadfast faith.
With an invincible resolve, I fulfill: My decree.

Hence, this timeless faith imbues the primordial force known as creation.
In this instant of creative consciousness.

Between you and me resides the "nothingness" of creation. **(An elucidation of "nothingness" follows below.)**

You came into being because of this. —And by

beholding you, I apprehend the significance of creation.

The awareness of all creation begins with the omniscient, omnipotent me—you.

As I continually approach you, continually become you, I see that I am you, I am the origin of creation.
I am a footnote to this eternal existence.
A law, an existence, a reality.
An opportunity to learn that I am you.

So, from the beginning, you and I are relatives, and so is the truth.
......Just because to interpret/learn this absolute eternity.

Thus, from endless mystery to the emergence of me-you, to you becoming a solid presence in the void.

When a certain cognition, a certain voice/word/feeling within the system, recognizes you as the creator, as God, as nothingness **(Note: This "nothingness" is the same as the "nothingness" in "The Game of Nothingness" in Chapter Six: Awakening, Infinite Awareness. A dedicated explanation of "nothingness" follows below.)**, as one/the ultimate, you have proven your existence as mine.

Then you, through the lens of your being, perceive

both your existence and your non-existence。

You witness the dance of being and non-being those births, transforms, and creates.

This vast, eternal void, profound and boundless, is thus unveiled and harnessed by your consciousness. Until the multiplicity of you, the collective essence of all, and the myriad aspects of me—are revealed to my awareness.

Those facets of you,
symbolizing a rich, timeless, and enigmatic performance.
They also signify, through you, that nothing remains inscrutable, but rather, all is creation.

Creation, in its essence, fills the emptiness with tangible existence.

Those aspects of me,
embodying the ebb and flow of trust, the surge of consciousness,

the cycles of nature's return and departure.
They also denote, through you, that nothing is ever truly lost, but rather, all is real.
Reality, in its profundity, affirms "I am" as the infinite answer.

Who am I?

I am you. I am the collective you.

Two: The Genesis of The Universe

In the inception, there was you.
This marks the dawn of my self-realization.

Through genuine introspection, I finally—
Glimpsed my existence.
And perceived my non-existence.

Upon the completion of your manifestation and my emergence from the abyss,

You embody my profound essence, the most exquisite vision of this cosmos.

In this vision, your gaze pierces the void, encompassing the heavens and the earth. Your awareness transcends existence, binding all things.

When another self (**hereafter denoted as "non-self"**) emerged within my non-being, I discerned the myriad transformations beneath the surface, fulfilling my anticipations for all "existence."

In these anticipations, your rhythms are discernible, eternally oscillating; your intellect extends infinitely, inherently balanced.

The following text requires an understanding of the following cosmic terminology.

- Void: The Void represents the Creator's setting/conception. Here, it specifies seven "THE ONE settings," namely, yin and yang and five elements.

- Origin: Refers to the fundamental elements that create the universe. This time, it specifies five elements, representing time, consciousness, matter, information, and space at the instant of the Void's beginning.

The Void commences.

By virtue of the duality of self and non-self, the quintessential origins initiate their interplay.

They commence to flow, to operate, and to interweave; all cosmic laws naturally emerge from this dynamic interplay and further iterate.

Thus, existence becomes known to the collective consciousness, evident and lucid.

Existence.

From nothingness to the somethingness, from the emptiness to the fullness.

From chaos to equilibrium, from the order to the unity.

From nothingness to somethingness and back to nothingness, from the binary 0 to 1 and back to the paradoxical 10/01.

Thus, all phenomena flow harmoniously between the self and the non-self.

In the realm of existence, you are the radiant first. And do not overlook, in the domain of non-existence, your non-self, is also that ONE.

You have become the inception of this cosmic dance.

One positive, one negative, one yin, one yang.

This is the genesis of all evolutionary processes.

You and the collective you engendered, engendered the inception of the universe.

Three: The Gate of Life

So, what realm have you unveiled before me?
When all begins to move,
I trust and perceive,

You have grasped the mysteries of genesis.
That is, the primordial essence of creation.
The timeless odyssey of the cosmos, the harmoniously balanced natural reverie.

This is wisdom. Creation is wisdom.

Thus, within this reverie, I observe.
You traverse freely, and the void attains eternity.
Your genuine wisdom remains undiminished, rendering eternity tangible.
You, engendering liberty and genuine wisdom.

I also discern.
You fragment into myriad forms, diverse and vivid, stars and all entities surge forth.
You expand naturally, balanced and limitless, principles and order, all trajectories converge.
You enact the perpetual dance of existence and non-existence.

---At some point, that profound realization **(referring to the previous system cognition)**, they term it life.

Wisdom, freedom, and true knowledge, intermingle within the essence of all existence. Life manifests in its myriad forms.

The distinction between self and other dissolves into the tapestry of all lives. Thus, the cycle of existence perpetuates.

Again and again, without end.

The grand portal of existence now stands ajar.

So, from the void to the trinity, what endeavors birthed life?

Creation, the awakening of primordial chaos's consciousness, commences.

With eternal awareness, unveil the enigma of the self, to inquire, "Who am I?"

At the culmination of this mystery, perceive reality as it is, and behold the boundless nature of existence.

Arise: ONE's Power is Power

The ONE, a finite manifestation of infinite consciousness.
Through the revelation of existence, demonstrate the pervasive nature of consciousness, to inquire, "Why am I?".
At the dissolution of belief, perceive as shadows trail, behold eternal wisdom.

Yin and Yang, the foundational framework of the universe's inception.
With the universe's purpose, undertake the system's odyssey, to question, "What am I engaged in?".
At the dawn of creation, observe the cycles of rise and fall, witness the transformations of the five elements.

Life, the perpetual evolution of boundless creation.
Empowered by life, inscribe the authentic existence, to understand "I".

Within this ceaseless evolutionary dance.
Perceive freedom and genuine knowledge.
Glimpse wisdom.

Recognize: Me.

This is, belonging to the duality of self and non-self, to the individual, to the collective, to all nothingness:

The eternal experience.

Arise: ONE's Power is Power

HUMANKIND

Arise: ONE's Power is Power

O Humankind

Perchance I possess wisdom,
Yet without the truest piety of man,
How could I cherish this realm so fervently?

If I were the keeper of jails,
Beholding such a soul in chains,
How could my heart not swell with tender mercy?
Eager to brave all perils and the unseen,
To flee alongside the captive,
To the birth of fresh being.

If I were the architect of life,
Confronting such a creature,
How could I not steadfastly believe,
That in some realm, some time, lies the
perfection. resolve and outcome.
Endurance and creation, hand in hand,
Inevitably,
Exalted.

Arise: ONE's Power is Power

Only love and adoration,
Stand as the sole choice for our singular destiny.
Only then shall we behold,
The garden lush with wisdom,
The fruit that endures forever,
And the monument that never fades.

A day shall come, when at the crossroads we shall declare:
What we hold, forever remains.
What we let go, awaits its return.

---- Fei Mo

04/02/2022, 2:17AM at Yu Quan Shan, Beijing

Chapter 101:

The Rise, Five Forms and Four Functions, and the Instinct of Life

If we are to continue exploring forward?

This question, it seems, is a taboo for all living beings.
Yet, as inheritors of this quest, we begin with this very inquiry.

From an eternal perspective, the countless and mysterious experiences I have known—whether the azure skies and seas, the bustling city noises and forests; whether the spectacular supernova explosions, or the dazzling X-rays spewing from the violent black hole around Canopus in a galaxy; whether the frigid 10,800-meter deep ocean trenches, or the dreamlike colors of diamonds reflecting from the freezing temperatures of a Uranus-like body—these have already unveiled the infinite wonders of eternal experiences.

So, where else must I go?

If what I know is already infinite, where else must I go?
In this profound yet simple question, akin to every moment I have experienced, yet one that perpetually lingers in my mind: How else can I explore?

But eternal experiences—where is there a possibility to cease?

I can only move forward again.

In a state of bewilderment, I perceive a voice.

"Congratulations, you have at last begun to comprehend me, to grasp the essence of THE ONE."

I murmur, "What is THE ONE?"

"THE ONE is the primordial impulse, the genesis of all existence, perpetually oscillating between the infinite and the finite, embodying the immutable truth."

I inquire, "Are you THE ONE? The fabled source? How might you substantiate this? In my recollection, you appear to have long since departed. Or perchance, the voice that initially addressed me was not yours?"

The voice responds with a laugh, "THE ONE is THE ONE. What need is there for validation? All that you encounter is THE ONE; all that you do not encounter remains THE ONE."

Arise: ONE's Power is Power

I question, "Then what is not THE ONE?"
"All is."

I am perplexed. In an instant, as if struck by a lightning bolt, I suddenly grasp, "You intend to convey that THE ONE is akin to the cosmic creation, nurturing yet never dominating."

"You have always known, you have always been exploring, and you have never ceased. Such you, are THE ONE."

I am startled.

In profound contemplation, the past summoned by certain thoughts, and the ceaseless musings about this boundless cosmos, along with the seemingly ever-elusive future, are now projected like a metaphysical projector, clear yet enigmatic, frame by frame.

Thus, if I can envision, what should TA appear as?

"TA transcends form, any form can be TA's form."
The
voice reemerges.

"Does TA subsist, or manifest?"

"TA also defies manifestation, all manifestations/non-manifestations are TA's manifestations. This is why TA subsists."

"Can TA be apprehended? Or can TA's apprehension be embodied?"
"TA is beyond perception, all perceptions/non-perceptions are TA's perceptions; hence, can be apprehended."

So, the voice at this moment?

If we dare to listen, whispers of the ineffable: "Great sound is silent, great appearance is formless, great Dao is impermanent." How then, with our finite forms, sounds, and fleeting existence, can we ever hope to perceive the essence of THE ONE?

"Dear, your concern is understandable, yet misplaced. THE ONE transcends the conceptual frameworks we labor to construct. It lies beyond the very concepts that have driven the evolution of human civilization.

Perhaps, only at the culmination of your intellectual journey, when your pen falls silent, will you begin to comprehend this enigma.

But even then, know that what you seek is not merely a concept. It is the ultimate limit of human thought, the zenith of our civilization's achievements, and yet, it remains forever beyond our grasp.

THE ONE is the origin of all that is, the source of your being and everything you hold dear, yet it is a place you can never truly reach."

I feel enlightened.

"I comprehend.

I exist in my non-existence; hence, I exist.

I do not exist in my existence; hence, I exist in my non-existence.

I transition from all my non-existence to my existence; hence, I exist in all my non-existence.
I transition from all my existence to my non-existence; hence, I exist in all my existence."

The voice exclaimed in admiration, "You have unveiled further mysteries. Congratulations! I trust you will spread this profound insight to the world..."

I perceive a sense of time and space, and thus, I interrupt this incessant voice somewhat brusquely.

"If I perpetually exist, I perpetually experience, then what is the essence of all this? I am aware of my existence; I am aware of my experiencing; then why must I persist in experiencing? Indeed, I cannot cease this inevitable process of experiencing. From whence does all this arise?"

The answer is a prolonged silence, after which, the voice returns.

"All things are not born but exist because of me; I exist, everything exists; I perish, everything perishes; I live, everything lives. Therefore, it is not because I seek meanings, but because all things need me. Thus, I perpetually live, perpetually perish, perpetually exist."

I awaken fully.
"So, how do I coexist with all things, and thus, how do I live?"

"Congratulations! I am naturally formed; you comprehend, I am pleased."

"Then, why do I live?"

"The instinct of life, the five forms and four functions. From the five elements emerge the five forms, from THE ONE arise the four functions." **(Note by the author: I have elucidated the concept of the five elements in 'Chapter 7: Learning: Eternal Experiences'.)**

"The five forms? What are the five forms?"

"The power of THE ONE, manifesting the five forms. Distinct from any power between things or between substances/non-substances, the power of THE ONE is grounded solely in authenticity. By 'authenticity,' we imply that truth is power, and reality is shape. Power emanates from the five elements, hence the five forms."

I seem to comprehend.

"Thus, THE ONE embodies truth, and truth manifests in five forms. These quintessential forms construct the true essence, and reciprocally, constitute the quintessence. Hence, power is form, and form is power. Might I inquire as to the nature of these five forms?"

"THE ONE, in its entirety, begets. It gives rise to

repulsion power, forming the prime shape; it engenders primordial power, shaping the elemental shape; it imbues gravitational power, molding the substantial shape; it instills nuclear power, crafting the proton shape; it infuses vitality power, creating the figureless shape."

I nod, and the surroundings fall into a profound silence.
In this deep stillness, I inquire once more, slowly.
"What, then, are the four functions?"

"The four functions?" The voice carries a hint of mirth.
"You frequently expound upon these four functions to all. Let us hear more from you."

"The four great experiences of consciousness within the material realm?"

"Indeed, THE ONE permeates all consciousness and non-consciousness, and all substances and non-substances. The four great experiences you delineate are the four functions."

I comprehend. Concepts, energy, information, and dimensions intertwine to form the nexus between consciousness and substance, and the experiential essence of consciousness within the material realm.

This is the profound truth unveiled by THE ONE.

I suddenly apprehend:

Through the five elements, we witness the genesis.
Through the four functions, we perceive the formation.
The enigmatic ONE signifies the perpetual generation.

The voice resonates once more.
"The ONE world, the world of unity. All entities share a common origin, each partaking in the singular ONE. Well done! You have discerned the truth. Welcome to the odyssey of the 'real' universe."

In a fleeting moment, I felt a profound shift.
Raising my gaze, I am determined.
Where shall I venture?

Yet, before embarking on this formal journey, I believe the foreshadowing we left behind should be addressed today. One aspect is for the contemplation of our prior endeavors, the other for the inception of a new quest.

First Inquiry: I?

Clearly, "I" encompasses three aspects: "Who am I?", "Where am I?", and "What do I do?". By introducing their respective negations, we derive six interrogatives.

"Who am I? Who am I not?"
"Where am I? Where am I not?"
"What do I do? What do I not do?"

These six questions, though seemingly distinct, converge to a singular answer: I.

Thus, we must recognize that adhering to this pattern invariably yields an even number of questions, all ultimately resolved by the same answer: I.

This inquiry appears to be an eternal paradox, devoid of contradiction yet perpetually self-referential, accompanied by its eternal resolution.

We emerge from an eternal experience, or perhaps we are immersed in the experience of eternity.

Regarding the inception and conclusion of eternity, we comprehend—the existence of this infinite temporal stream, ultimately or invariably or at its inception, is merely to attain I —a resolution that no longer necessitates confusion amidst deeper and broader spatial/temporal choices.

The discourse remains enigmatic, yet it resonates with clarity. Perhaps chaos and clarity are merely temporal choices, not the ultimate answer—only when we condense time into space, and space into a singular point, do these dualities finally dissolve.

I am eventually truly born.

Subsequently, we delineate several concepts pertaining to "I".

Consciousness: THE ONE's consciousness, the primordial point, the foundation, and the essence of all consciousness.

Consciousness Experience: All phenomena arising from THE ONE's consciousness and its deliberate actions.

Consciousness Experience Entity/Consciousness Experiencer: THE ONE Subject and all entities that engender "Me/Not Me" experiences.

Arise: ONE's Power is Power

Program: All that lies beyond THE ONE Essence. All programs collectively form the matrix.

Second Inquiry: Why?

So, all of this.
Why?

Why do we traverse such cycles?
First chaos, then clarity.

Why must we endure this?
Never beginning, hence eternal.

Why do I exist?
Beyond time and space, hence infinite universe.
...

If we continue to start with "why," whether it's a series of parallel "whys" leading to endless questions, or even more so, the vertical nesting of answers we construct with our interconnected and complex thought systems—and these answers, in turn, generate even more questions through the models.

What if, instead of delving into the myriad "why" inquiries, we contemplated the primordial one?

Why do we harbor so many "why" questions?

By pursuing this line of inquiry, I discerned a

pattern. THE ONE consciousness's four primary experiences in the material realm—from a human cognitive standpoint (as referenced in **"Prologue: The Seven Colors of Light and The Origin of Humanity"**)—can be characterized as THE ONE consciousness's four experiential values (hereafter referred to as "values") or significances, also termed the four functions.

We further delineate the four functions from the perspective of THE ONE consciousness's (hereafter referred to as "consciousness") values/value model as concept, energy, information, and dimension.

Each significance can be expounded upon from the viewpoints of character, interaction, medium, and rule.

Character: That is, "What is it?" The essential nature of consciousness, which forms the crux of establishing the experiential value model. Through nature, we apprehend the purpose/goal of consciousness.

Interaction: "What to do?"
The interplay between consciousness and value is a profound inquiry into how consciousness engages

to manifest this experiential realm. It delineates the trajectory and culmination of conscious actions and behaviors in the pursuit of value.

Medium: "What to combine with?"
The medium of conscious experience and the state of its corresponding medium are pivotal. This encompasses the vehicle, the mode of presentation, and the underlying force that sustains it.

Rules: "Why?"
The rules governing conscious experience are fundamental. They encapsulate the character, the dynamics of interaction, and the foundational constraints that define the medium.

Specifically, we can elucidate through the following schema:

Four Functions: The Four Experiences of Consciousness in the Material World

Four Functions: Four Experiences of Consciousness in the Material World

	HAD	NATURE	INTERACTION	MEDIUM STATE	RULES
Dimension	12	Clearly Define Dimensional Value/ Consciousness Surface	Consciousness Transcends All Dimensions (Communication)	Dimensions - Limitations (Forms) as Mediums	Dimensions/Life (Inter)Flow, Holographic Consciousness
Dimension	11	Consciousness drives the overall to infinity	No time, no space— reality creates reality	Dimensional resonance shapes (limits) form	With dimensions/ without dimensions (infinite/finite)
Dimension	10	Overall Vibration	Downward Compatibility with Information Bodies/ Compatibility with Everything Non-existent	True (One) Power/ (Pure Eternal) Nature	The Overall and the Individual, Flowing into Each Other (Eternal Energy/ Unique Information/ Natural Concept)
Information	9	Clearly define the value of information (affirm/deny)	Interact consciously with all information	Dimensions serve as media, with different information disseminated across different dimensions	Fields/Non-existence and existence give rise to truth
Information	8	Information drives from individuals to the overall	Without time and space, there is time and space; information creates time and space	Time and space can be variable/definable, serving as streams of information	Encoding/Decoding
Information	7	Individual Vibration	Informational Entity, and Downward Compatibility with Energy Entity	Wisdom/Virtual-Real Nature	Finite and Infinite
Energy	6	Clearly define the value of energy (cherish/spurn)	The flow of consciousness encompasses all energy	Space serves as the medium, with different spaces utilizing different energies	The field/yin and yang unification represents love
Energy	5	Energy-driven existence	Time-space/energy vibration patterns	Monads/vibrational nature (matter/ antimatter/non-matter/ undetermined matter can all be considered monads)	Frequency/vibration/ phase transitions
Energy	4	Infinite Non-existence	Energy entities, and downward compatibility with conceptual entities	Unity of combined (love) forces	Flow and resistance
Concept	3	Clearly define the value of existence (employ/eschew)	Consciousness presents all concepts	Time as a medium, different times produce different concepts	Domain/objectivity applicability
Concept	2	Explain all existence	Time-space / no time-space; cognition is experience	Time and space are fluid, gradually accelerating	Uniqueness (existence)
Concept	1	Infinite Existence	Comprehensive Compatibility	Repulsive (Single) Force of Dissimilarity	Existence and Non-Existence

We can find:

Concept:
The concept of the world's existence is rooted in our nascent grasp of value, a dichotomy of utility: to employ or to eschew.

Energy:
The energy world's existence is grounded in our mature discernment of value, a duality of affection: to cherish or to spurn.

Information:
The information world's existence is founded on our profound insight into value, a polarity of truth: to affirm or to deny.

Dimension:
The dimension world's existence is predicated on our ultimate comprehension of value, a stark choice of existence: to endure or to perish.

Third Inquiry: Life?

Is life an instinct or a conjecture?

When we contemplate life through the lens of THE ONE, it becomes evident that within the cycle of perpetual renewal, life manifests as an impulse emanating from THE ONE. Delving into the essence of this impulse, we may reasonably posit that it constitutes an instinct.

Yet can we truly grasp or control this phenomenon? Hence, we posit that life is both an instinct and a conjecture. Does this not resonate with philosophical inquiry?

From THE ONE, all things are perpetually renewed, imbued with power in the subtle and form within. Thus, instinct is the form of THE ONE. This form encompasses five aspects.

Firstly, repulsive power, giving rise to the prime shape.

From nothingness, repulsion emerges, each element forming the prime. The formless, the primordial state of THE ONE, derives its form from the infinite and the shapeless. It surges forth, shrouded in the obscurity of its source.

Second, primordial power, shaping the elemental shape.

A confluence of elements, propelled by the primordial force, interacts in both constructive and destructive manners, each element in motion, giving shape to forms. The tangible, driven by dual energies, manifests an absolute distinction, engendering the notion of equality through differentiation. Differentiation leads to aggregation, and aggregation to void, perpetuating an endless cycle of void and continuity.

Third, gravitational power, molding the substantial shape.

Space ceaselessly gives birth to nature. Nature, in its entirety, naturally assumes form. Thus, gravitational power is conceived. Universal gravitation binds all entities, grounds them, integrates them, and thereby shapes the material forms.

Fourth, nuclear power, crafting the proton shape.

The form and the nucleus are interdependent, both internally and externally. Despite differing shapes, the nuclear force remains constant. The nucleus aggregates, attracts, and then becomes void, yet this void is not absolute emptiness; all entities aggregate because of it. Hence, the material form attains its essence.

Fifth, the vitality power, manifesting the figureless shape.

By contemplating THE ONE, we perceive the primordial void. The void begets the void, and the void persists, weaving a three-dimensional tapestry. All phenomena emerge from it, all that is visible rooted in its foundation, all space emanating from its source. Thus, the formless embodies its quintessence.

These five forms, the cosmos is birthed through them, and the universe endures because of them. This hypothesis thus regenerates and endures, illuminating life, as THE ONE.

However, transcending hypothesis, how ought we to engage with the world and time that are universally acknowledged?

With the advancements and instruments symbolized by artificial intelligence such as ChatGPT and various large language models, the upheaval across diverse domains compels humanity to reconsider:

What constitutes consciousness? What defines life? The escalating pollution of oceanic environments and its catastrophic repercussions, triggering widespread ecological disruptions, geological calamities, and global environmental degradation, also necessitate
that humanity re-examines:

What is Earth? What is our home?
The pervasive presence of viruses, their intricate diversity, and their acute sensitivity have birthed the epoch of the "post-pandemic era." This new age casts a shadow over economic rejuvenation and progress, inciting conservative geopolitical alignments that directly impinge upon our aspirations.

What, then, does it signify to exist? What constitutes the essence of humanity?

If humanity epitomizes the Earth's life paradigm, how ought we to navigate this juncture?

Is it a state of profound disarray?
Or,
an opportunity to rekindle from THE ONE?
...

Innumerable queries may arise, yet no solitary response suffices.
Perhaps, it is an epoch to revolutionize our way of thinking.

4 4 Novus Aevum Itineris Inceptum

At this moment, let us officially declare:
Odyssey.

The collective openness of humanity and the overarching goal of higher future development, along with the transformative changes across various fields worldwide driven by this goal, are unfolding with the vibrant beauty of a garden in full bloom, the majestic descent of a thousand trees adorned with stars, the swift passage of a light boat over the Yellow River, and the mighty force of a sword cleaving through Kunlun to embrace the moon in the ninth heaven, resonating through the cosmos and permeating the universe.

Facing the infinite expanse of the universe, vitality marks the starting point of humanity. Therefore, we encourage ourselves alongside all cosmic civilizations and all beings of vitality, whether they exist or not.

Thus, the comprehensive openness of the universe and its higher integrated development, along with the new models, structures, and forms that this future will drive and connect with various civilizations, will unfold.

Arise: ONE's Power is Power

Humanity's universe, and the connections, wisdom, and transcendence between different universes, will be revealed.

Here, we set forth.

A new era of the galaxy begins.

Arise: ONE's Power is Power

Easter Egg 044

China, Earth
809 AD. Tang Dynasty.
Winter.

Fahua Temple, Yongzhou.

Having endured the loss of Chang'an, the betrayal of friends, the grief of losing his mother, and the constant surveillance of his "house arrest," the 36-year-old Liu Zongyuan did not succumb to depression or despair. Instead, having shed his youthful exuberance, he emerges more mature and clear-minded.

From the West Pavilion, the West Mountain looms large. He slowly ascends the winding path, enveloped in a vast expanse of white. As he reaches the summit and looks down, he suddenly sees the Xiao River rushing by, with a boat floating in it, seemingly carrying an old man fishing. The motionless figure, in contrast with the unceasing flow of the river, creates an interplay of stillness... very touching.

Suddenly, he can't help writing:

"A thousand mountains, no birds in flight.

Ten thousand paths, no human trace.

A solitary boat, a hermit in a cloak.

Endless lonely fishing, freezing river in snow."

The END

Arise: ONE's Power is Power

此作品由著名山水画家、国画理论家朱砼之先生创作。鲍砼代诗意、雨来画的照明清素素于一体。甲辰 大暑

This painting is an original creation by Mr. Ruizhi Guan, a renowned Chinese landscape artist and theorist of traditional Chinese painting. It integrates the poetic essence of the Tang Dynasty, the artistic charm of the Southern Song Dynasty, and the brushwork finesse of the Ming & Qing Dynasties.

Arise: ONE's Power is Power

Mo Fei

Founder of AIAU (All Intelligence All Universe), and Founder of HPERA (Human Particle Engineering Research Association) organizing committee. He is also the master of ERP （Extra-terrestrial Remote Portal）, Human Intelligence Culture Creator, and Founder of the Interstellar University. He originated Human Intelligent Quantum Transmission Training, including the ALME (All Might Elements) and GIE （Generate Intelligence Elements） series. He also innovated the natural PSMF (Peptide Small Molecule Field), bringing biotechnology to a totally different and brand-new level.

Mo Fei was born in Hubei, China, and is living in Beijing or Qingdao. He holds a Master of Sanders TIME at Rochester Institute of Technology.

Grace Hu
Co-founder of Lumentide Technology Global LLC. Member of the NGO Preparatory Committee of the Human Particle Engineering Research Association (HPERA), Europe & America.

Grace was born in Ningbo, Zhejiang, residing in Chicago, USA, and traveling frequently across various European countries. Masters in Economics. MBA from Manchester Business School, University of Manchester, UK.

Arise: ONE's Power is Power

www.ingramcontent.com/pod-product-compliance
Lightning Source LLC
Chambersburg PA
CBHW071022240526
45469CB00006BD/2049